An Account Of Sa-Go-Ye-Wat-Ha Or Red Jacket And His People, 1750-1830

By

John N. Hubbard

Double 9 BOOKS

An Account Of Sa-Go-Ye-Wat-Ha Or Red Jacket And His People, 1750-1830

by John N. Hubbard

ISBN: 978-93-59950-07-5

Published by

DOUBLE 9 BOOKS

2/13-B, Ansari Road
Daryaganj, New Delhi – 110002
info@double9books.com
www.double9books.com
Tel. 011-40042856

ABOUT THE AUTHOR

John N. Hubbard is a distinguished creator and historian acknowledged for his substantial research and writings, in particular within the realm of Native American history and the early interactions between Indigenous peoples and European settlers. One of his most first rate works is "An Account of Sa-cross-ye-wat-ha or Red Jacket and His People, 1750-1830." In this book, Hubbard takes readers on a charming adventure into the existence and instances of Sa-go-ye-wat-ha, a prominent Seneca chief called Red Jacket. Set towards the backdrop of the 18th and early nineteenth centuries, the narrative paints a shiny photo of the Seneca humans and their reports at some stage in a length of great trade, marked by using European colonization, American Revolution, and the enlargement of america. Hubbard's paintings masterfully combine factors of biography and ancient documentation, providing a complete account of Red Jacket's lifestyles, management, and the demanding situations confronted with the aid of the Seneca Nation. It delves into Red Jacket's efforts to navigate the turbulent waters of international relations and war even as striving to defend his humans's rights and subculture. This e book is a testament to Hubbard's willpower to preserving the records and cultural historical past of Indigenous communities.

CONTENTS

PREFACE

The "Life and Times of Red Jacket" by Colonel William L. Stone, has been before the public for many years. The industry and ability of the author have made it a work of great value, and his extensive researches have left but little room for anything new to be said, by one coming after him. Yet the fact need not be concealed that many, who were intimately acquainted with Red Jacket, were disappointed when they came to read his biography. If it had been prepared under the direct influence and superintendence of Thayendanegea, or Brant, it could not have reflected more truly the animus of that distinguished character. Red Jacket in his day was the subject, at different times of much angry feeling, and jealousy. The author has not taken pains to embalm it, in these memorials of the great orator of the Senecas. Much that was the subject of criticism during his life, admits of a more charitable construction, and the grave should become the receptacle of all human resentments.

The author acknowledges his indebtedness to the labors of Col. Stone, and by an honorable arrangement, liberty was obtained for the use made of them, in the following pages. Acknowledgments are due also to others, whose names will appear in the course of this work.

TRACY, CAL., *April 12th*, 1885.

CHAPTER I

Name widely known—Interest naturally awakened in his history—His origin —Development of his genius—Opinion of Capt. Horatio Jones—Customs of his people—Their love of eloquence—Distinguished orators among them—The inviting field opened.

Among the aborigines of this country, few names have excited a deeper interest, or have been more widely and familiarly known than that of RED JACKET. The occasion of this notoriety was the rare fact that, though a rude and unlettered son of the forest, he was distinguished for the arts and accomplishments of the orator. His life marks an era in the history of his nation and his name like that of Demosthenes, is forever associated with eloquence.

Other circumstances however, impart interest to his history. His was the last great name of a nation, and he is entitled to remembrance, on the soil which was once the home of his fathers. And though linked with a melancholy association, as connected with the waning history of a people that once laid a claim to greatness, but are now fast passing into obscurity, it is not on this account the less attractive, but presents another reason for our regard.

Such was the name of SA-GO-YE-WAT-HA, or, as he has more commonly been called, Red Jacket. Having risen, by the force of his eloquence, from an obscure station to the highest rank among his people, he became conspicuous in all of those great transactions, in which they gradually relinquished a title to their old hunting grounds, and gave place to the intrusive white man. And he lived to see his nation pass from the pride of their ancient dominion, to so humble an inheritance, that his last days were embittered with the thought, that the *red men* were destined to become extinct. With him has ceased the glory of their council fire, and of their name.

His origin, as we have intimated, was obscure. He must be introduced, as he has come down to us, without rank or pedigree. His pedigree nature acknowledged, and gave him a right to become great among her sons. His birth is a matter of fact, its time and place, circumstances of conjecture. Some affirm that he was born at the Old Seneca Castle, near the foot of

Seneca lake, not far from 1750. [Footnote: Hist. of North American tribes by Thos. L. McKenney.]

Another tradition awards the honor of his birth to a place at, or near Canoga, on the banks of the Cayuga lake. [Footnote: Schoolcraft's Report.]

Who were his parents? and what, his early history? As the wave casts upon the shore some treasured fragment, and then recedes to mingle with its parent waters, so their names, and much of his early history have been lost in the oblivion of the past.

So likewise it is uncertain, as to the time when the wonderful powers of his genius began to be developed, or as to the steps by which he arrived at the high distinction of orator among his people.

Whether by dint of study he gained the requisite discipline of mind, and acquired that elegance of diction for which he was distinguished; whether by repeated trial and failure, accompanied by a proud ambition, and an unyielding purpose, he reached, like Demosthenes, the summit of his aspirations; or, assisted more by nature than by art, emerged, like Patrick Henry at once, into the grand arena of mind, and by a single effort attained distinction and fame, is to be gathered more from circumstances than from facts.

It is generally conceded, however, that the powers of his intellect were of the highest order. Captain Horatio Jones, the well known interpreter and agent among the Indians, and than whom no one was more intimately acquainted with this orator of the Seneca nation, was accustomed to speak of him as the greatest man that ever lived. "For," said he, "the great men of our own and of other times, have become so by education; but RED JACKET WAS AS NATURE MADE HIM. Had he enjoyed their advantages, he would have surpassed them, since it can hardly be supposed that they, without these, would have equalled him." [Footnote: Conversation of the author with Col. Wm. Jones, of Geneseo, Livingston Co., N. Y., son of Capt. Horatio Jones.]

Some allowance should be made for this statement, perhaps, on the ground that Mr. Jones was a warm admirer of the orator's genius; yet his admiration sprang from an intimate knowledge of him, seen under circumstances, that afforded the best opportunity of forming a just opinion of his talents; and these, he maintained, *"were among the noblest that nature ever conferred upon man."*

But genius, while it may have smoothed the way, may not have spared him the pains, by which ordinary minds ascend to greatness. For since it is so universally the fact, that the path to eminence, is rugged and steep, and the gifts of fame seldom bestowed but in answer to repeated toil; curiosity would inquire by what means one, who was reputed a barbarian, gained the highest distinction ever awarded to civilized man. It is not enough to reply simply, *"that nature made him so,"* or to receive, without qualification, his own proud assertion, "I AM AN ORATOR, I WAS BORN AN ORATOR." The laws of mind are the same for peasants, and princes in intellect; great minds as well as small, must take measures to compass their object, or leave it unattained.

It does not appear that his genius was sudden, or precocious in its development. It is said that his mind, naturally active and brilliant, gradually opened, until it reached its meridian splendor. Nor did his powers grow without any means to mature and perfect them. As the young oak is strengthened by warring with the storm, so the faculties of his mind gained force by entering freely into conflicts of opinion. Accustomed to canvass in private the questions which agitated the councils of his nation, he began to ascertain the reality of his own power, and by measuring his own with other minds, he gained the confidence that flows from superior wisdom. [Footnote: Conversation with Col. Wm. Jones.]

The tastes and regulations of his own people favored very much, the promptings of his genius. They were lovers of eloquence, and their form of government fostered its cultivation. This though differing but little from the simplicity found in rude states of society, presented a feature peculiar among a people not far advanced in civilization, which served greatly to promote elevation of mind, and advance them far above a condition of barbarism. They were in the habit of meeting in public assemblies, to discuss those questions that pertained to the interests, or destiny of their nation. Around their council fires their chiefs and warriors gathered, and entered freely, so far as their dignity, consideration, or power of debate admitted, into a deliberation on public affairs. And here were manifested an ability and decorum which civilized nations even, have viewed with admiration and surprise. For though we might suppose their eloquence must have partaken of rant and rhapsody, presenting a mass of incoherent ideas, depending for their interest on the animation of gesture and voice, with which they were uttered, yet we would do injustice to their memory, if we did not give their orators the credit of speaking as much to the purpose, and of exhibiting as great a force of intellect, as many who would claim a higher place than they in the scale of intelligence and refinement.

Many of their orators were distinguished for strength of mind, and in native power of genius, might compare favorably, with the men of any age or clime. The names of Garangula, Adario, Hendrick, Skenandoah, Logan and others, might be mentioned with pride by any people.

GARANGULA, has been styled the very Nestor of his nation, whose powers of mind would not suffer in comparison with a Roman, or more modern Senator. [Footnote: Drake.]

ADARIO is said to have been a man of "great mind, the bravest of the brave," and possessing altogether the best qualities of any Indian known to the French in Canada. [Footnote: Charlevoix.]

It has been remarked of HENDRICK, that for capacity, bravery, vigor of mind, and immovable integrity united, he excelled all the aboriginal inhabitants of the United States, of whom any knowledge has come down to the present time. [Footnote: Dr. Dwight.]

SKENANDOAH in his youth was a brave and intrepid warrior, and in his riper years one of the best of counsellors among the North American tribes. He possessed a most vigorous mind, and was alike active, sagacious, and persevering. He will long be remembered for a saying of his to one who visited him toward the close of life; "I am," said he, "an aged hemlock, the winds of an hundred winters have whistled through my branches. I am dead at the top. The generation to which I belonged has run away and left me." He was a sincere believer in the Christian religion, and added to the above "why I live the Great and Good Spirit only knows. Pray to my Jesus, that I may have patience to wait for my appointed time to die." [Footnote: Annals of Tryon County.]

And Mr. Jefferson regarded the appeal of LOGAN to the white men, after the extirpation of his family, as without a parallel in the history of eloquence.

These were men who have been revered by the civilized world, as worthy of a place with the distinguished and great among mankind.

"Oratory was not alone a natural gift, but an art among the Iroquois. It enjoined painful study, unremitting practice, and sedulous observation of the style, and methods of the best masters. Red Jacket did not rely upon his native powers alone, but cultivated the art with the same assiduity that characterized the great Athenian orator. The Iroquois, as their earliest English historian observed, cultivated an Attic or classic elegance of speech, which entranced every ear, among their red auditory." [Footnote: Mr. Bryant's speech.]

Those public games, entertainments, religious ceremonies and dances, common among the Indian tribes, added interest to their council gatherings, and made them a scene of attraction for the entire nation. Thither the young and old of both sexes were accustomed to resort, and, assembled at their national forum, listened with profound attention and silence to each word spoken by their orators. "The unvarying courtesy, sobriety and dignity of their convocations led one of their learned Jesuit historians to liken them to the Roman Senate." [Footnote: W. C. Bryant's speech before the Buffalo Historical Society on the occasion of the re- interment of Red Jacket's remains.]

"Their language was flexible and sonorous, the sense largely depending upon inflection, copious in vowel sounds, abounding in metaphor; affording constant opportunity for the ingenious combination and construction of words to image delicate, and varying shades of thought, and to express vehement manifestations of passion; admitting of greater and more sudden variations in pitch, than is permissable in English oratory, and encouraging pantomimic gesture, for greater force and effect. In other words it was not a cold, artificial, mechanical medium for the expression of thought or emotion, or the concealment of either, but was constructed, as we may fancy, much as was the tuneful tongue spoken by our first parents, who stood in even closer relations to nature." [Footnote: Ib.]

Hence, though the Iroquois were a warlike people, and delighted in deeds of bravery, there was an inviting field opened to one, who could chain their attention by his eloquence, and sway their emotions at will.

Such advantages being presented for the exercise of the powers of oratory, it can hardly be supposed that a mind endowed as richly, as was Red Jacket's, by the gifts of nature, would fail to perceive the path in which lay the true road to eminence among his people. And his subsequent career indicates but too clearly, the choice he made of the field in which to exercise his noble powers.

CHAPTER II

Glance at the early history of the Iroquois—The territory occupied—Tuscaroras—Original strength—Traditions—Probable course of migration—The Senecas—Story of their origin—Singular romance.

Rising up from the obscurity of the past, we find a people, singular in their habits and character, whose history has been strangely, and in some respects sadly interwoven with our own. They were the original occupants of the soil, claiming to have lived here always, and to have grown out of the soil like the trees of the forest. Scattered over this continent were various Indian tribes, resembling each other in their general features and habits, but in some instances exhibiting stronger and more interesting traits of character than the others. Among these were the Iroquois, and if Red Jacket was distinguished among his own people, his own people were not less conspicuous among the North American Indians.

He sprang from the Senecas, and was accustomed to speak of his origin with feelings of conscious pride. For the Senecas were the most numerous and powerful of the six nations, of whom they were a part. Such was the title given to that celebrated Indian confederacy which, for a length of time unknown to us, inhabited the territory embraced by the State of New York.

Here they lived in a line of settlements extending from one end to the other, through the middle of the State, and their domain as thus occupied, they were accustomed to style their *Long House*. It was a shadowy dome, of generous amplitude, covered by the azure expanse above, garnished with hills, lakes, and laughing streams, and well stored with provisions, in the elk and deer that bounded freely through its forest halls, the moose that was mirrored in its waters, and the trout, those luscious speckled beauties, that nestled cosily in its crystal chambers.

The eastern door was guarded by the Mohawks, who resided at one, and its western by the Senecas, who dwelt at the other extremity of this abode.

When ever a messenger from another nation came to them on business, or knocked, as it was termed, at the eastern or western door of their *long house*, it was the duty of the nation to which he came, to give him entertainment, and examine into the nature of his embassy. If it was of small importance, it was decided by their own council; but if it was such as to demand the united wisdom of the tribes, a runner was sent with a belt of wampum to the nearest nation, which would take the belt and send a runner with it to the next, and so on, and thus with but little delay, a general meeting was summoned of all the tribes.

This confederacy at one time consisted of five nations, but afterward embraced six, by the addition of the Tuscaroras, a tribe that once occupied the territory of North Carolina.

This tribe is said to have belonged at an early day to the Iroquois family, and to have inherited the enterprising and warlike character of the parent stock. They fought successfully with the Catawbas, Cowetas, and the Cherokees, and thought to exterminate by one decisive blow, all of the white inhabitants within their borders. Unsuccessful in the attempt, pressed sorely by the whites, who resisted the attack, and unwilling themselves to submit, they removed to the north, and through sympathy, similarity of taste, manners, or language, or from the stronger motives of consanguinity, became incorporated with the confederated tribes of the Iroquois. [Footnote: Schoolcraft's Report. Mr. Schoolcraft prefers, and quite justly the name Iroquois, as descriptive of this confederacy, instead of Six Nations, since the term is well known, and applicable to them in every part of their history. Whereas the other is appropriate only during the time when they were numerically six.]

Thus constituted they presented the most formidable power, of which we have any knowledge in the annals of the Indian race. By their united strength they were able to repel invasion, from any of the surrounding nations, and by the force of their arms and their prowess in war, gained control over an extent of territory much greater than they occupied.

They sent their war parties in every direction. The tribes north, east, south, and west of them were made to feel the power of their arms, and yield successively to their dexterity and valor. Now they were launching their war-canoes upon the lakes and rivers of the west, now engaged in bloody conflicts with the Catawbas and Cherokees of the south, now traversing regions of snow in pursuit of the Algonquins of the north, and anon spreading consternation and dread among the tribes at the remotest east. Their energy and warlike prowess made them a terror to their foes, and distant nations pronounced their name with awe.

By what means these several tribes had been brought to unite themselves under one government, how long they had existed in this relation, and what was the origin of each one, or of all, are questions which will never perhaps be fully determined. There being no written records among them, all that can be ascertained of their history previous to their becoming known to the whites, must be gathered from the dim light of tradition, from their symbolic representations, from antique remains of their art, and from their legends and myths. These present in an obscure and shadowy form, a few materials of history, whose value is to be measured by the consideration, that they are all we have to tell the story of a noble and interesting race of men.

Their traditions speak of the creation of the world, the formation of man, and the destruction of the world by a deluge. They suppose the existence originally of two worlds, an upper and lower. The upper completed and filled with an intelligent order of beings, the lower unformed and chaotic, whose surface was covered with water, in which huge monsters careered, uncontrolled and wild. From the upper there descended to the lower a creating spirit, in the form of a beautiful woman. She alighted on the back of a huge tortoise, gave birth to a pair of male twins and expired. Thereupon the shell of the tortoise began to enlarge, and grew until it became a *"big island"* and formed this continent.

These two infant sons became, one the author of *good*, the other of *evil*. The creator of *good* formed whatever was praiseworthy and useful. From the head of his deceased mother he made the sun, from the remaining parts of her body, the moon and stars. When these were created the water- monsters were terrified by the light, and fled and hid themselves in the depths of the ocean. He diversified the earth by making rivers, seas and plains, covered it with animals, and filled it with productions beneficial to mankind. He then formed man and woman, put life into them, and called them Ong-we Hon-we *a real people*. [Footnote: This term is significant of true manhood. It implies that there was nothing of sham in their make up.]

The creator of *evil* was active in making mountains, precipices, waterfalls, reptiles, morasses, apes, and whatever was injurious to, or in mockery of mankind. He put the works of the *good* out of order, hid his animals in the earth, and destroyed things necessary for the sustenance of man. His conduct so awakened the displeasure of the *good*, as to bring them into personal conflict. Their time of combat, and arms were chosen, one selecting flag-roots, the other the horns of a deer. Two whole days they were engaged in unearthly combat; but finally the *Maker of Good*, who had chosen the horns of a deer, prevailed, and retired to the world above. The *Maker of Evil* sank below to a region of darkness, and became the *Evil Spirit*, or Kluneolux of the world of despair. [Footnote: Schoolcraft's Indian Cosmogony.]

Many of their accounts appear to be purely fabulous, but not more so perhaps than similar traditions, to be found in the history of almost every nation.

The Iroquois refer their origin to a point near Oswego Falls. They boldly affirm that their people were here taken from a subterranean vault, by the Divine Being, and conducted eastward along the river Ye-no-na-nat-che, *going around a mountain*, now the Mohawk, until they came to where it discharges into a great river running toward the mid-day sun, the Hudson, and went down this river and touched the bank of a *great water*, while the main body returned by the way they came, and as they proceeded westward, originated the different tribes composing their nation; and to each tribe was assigned the territory they occupied, when first discovered by the whites. [Footnote: Account by David Cusick, as contained in Schoolcraft's report. Mr. S. regards this account correct as indicating the probable course of their migrations.]

The Senecas, the fifth tribe of the Iroquois, were directed in their original location, to occupy a hill near the head of Canandaigua lake. This hill, called Ge-nun-de-wa, is venerated as the birth place of their nation. It was surrounded anciently by a rude fortification which formed their dwelling in time of peace, and served for a shelter from any sudden attack of a hostile tribe. Tradition hallows this spot on account of the following very remarkable occurrence.

Far back in the past, the inhabitants of the hill Genundewa, were surprised on awaking one morning, to behold themselves surrounded by an immense serpent. His dimensions were so vast as to enable him to coil himself completely around the fort. His head and tail came together at its gate. There he lay writhing and hissing, presenting a most menacing and hideous aspect. His jaws were widely extended, and he hissed so terribly no one ventured to approach near.

The inhabitants were thus effectually blockaded. Some endeavored, but in vain, to kill this savage monster. Others tried to escape, but his watchful eyes prevented their endeavors. Others again sought to climb over his body, but were unable; while others still attempted to pass by his head, but fell into his extended jaws. Their confinement grew every day more and more painful, and was rendered doubly annoying by the serpent's breath, which was very offensive.

Their situation drove them at length to an extremity not to be endured. They armed themselves with hatchets, and clubs, and whatever implements of war they could find, and made a vigorous sally upon their dreadful foe, but, alas! were all engulfed in his terrific jaws.

It so happened that two orphan children remained, after the destruction which befell the rest. They were directed by an oracle to make a bow of a certain kind of willow, and an arrow of the same, the point of which they were to dip in poison, and then shoot the monster, aiming so as to hit him under his scales.

In doing this, they encountered their adversary with entire success. For no sooner had the arrow penetrated his skin, than he presently began to grow sick, exhibiting signs of the deepest distress. He threw himself into every imaginable shape, and with wonderful contortions and agonizing pains, rolled his ponderous body down along the declivity of the mountain, uttering horrid noises as he went, prostrating trees in his course, and falling finally into the lake below.

Here he slaked his thirst, and showed signs of great distress, by dashing about furiously in the water. Soon he vomited up the heads of those whom he had swallowed, and immediately after expired and sank to rise no more. [Footnote: As related to the author by Col. Wm. Jones.]

From these two children, as thus preserved, the Seneca nation are said to have sprung.

So implicitly has this tradition been received by the Senecas, that it has been incorporated into the solemnities of their worship, and its remembrance continued from one generation to another by the aid of religious rites. Here they were formerly in the habit of assembling in council, and here their prayers and thanksgivings were offered to the Great Spirit, for having given them birth, and for rescuing their nation from entire destruction.

In speaking of this to the whites, they point to the barren hillside, as evincing the truth of the story, affirming that one day the forest trees stood thick upon it, but was stripped of them by the great serpent as he rolled down its declivity. The round stones found there in great abundance, resembling in size and shape the human head, are taken as additional proof, for they affirm that these are the heads disgorged by the serpent, and have been petrified by the waters of the lake. [Footnote: The author remembers well that in conversation with a Seneca Indian on this point, he seemed to take it as quite an affront that doubts should be expressed by the white people as to the reality of this occurrence.]

If nearness of locality will justify a glance of the eye for a moment, to an object not directly in the line of our pursuit, we might survey in passing a bold projecting height, not far from the hill Genundewa, marked by a legend which draws a tear from the eye of the dusky warrior, or sends him away in a thoughtful mood, with a shade of sadness upon his usually placid brow. The story is not of the same character and is of a more recent date than that of the serpent, but is said to be of great antiquity. It has been written with great beauty by Col. Stone, and as we are authorized, we present it in his own language.

"During the wars of the Senecas and Algonquins of the north, a chief of the latter was captured and carried to Genundewa, whereon a fortification, consisting of a square without bastions, and surrounded by palisades, was situated. The captive though young in years, was famed for his prowess in the forest conflict, and nature had been bountiful to his person in those gifts of strength and symmetry, which awaken savage admiration. After a short debate he was condemned to die on the following day, by the slow torture of empalement. While he was thus lying in the cabin of death, a lodge devoted to condemned prisoners, the daughter of the sachem brought him food, and struck with his manly form and heroic bearing, resolved to save him or share his fate. Her bold enterprise was favored by the uncertain light of the gray dawn, while the solitary sentinel, weary of his night-watch, and forgetful of his duty, was slumbering. Stealing with noiseless tread to the side of the young captive, she cut the thongs wherewith his limbs were bound, and besought him in breathless accents to follow her.

"The fugitives descended the hill by a wooded path conducting to the lake; but ere they reached the water, an alarm whoop, wild and shrill, was heard issuing from the waking guard. They tarried not, though thorny vines and fallen timber obstructed their way. At length they reached the smooth beach, and leaping into a canoe previously provided by the considerate damsel, they plied the paddle vigorously, steering for the opposite shore. Vain were their efforts. On the wind came cries of rage, and the quick tramp of savage warriors, bounding over rock and glen in fierce pursuit. The Algonquin with the reckless daring of a young brave, sent back a yell of defiance, and soon after the splash of oars was heard, and a dozen war canoes were cutting the billows in their rear. The unfortunate lovers on landing, took a trail leading in a western direction over the hills. The Algonquin, weakened by unhealed wounds, followed his active guide up the aclivity, with panting heart and flagging pace; while his enemies, with the grim old sachem at their head, drew nearer and nearer. At length finding further attempts at flight useless, she diverged from the trail, and conducted her lover to a table-crested rock that projected over a ravine or gulf, one hundred and fifty feet in depth, the bottom of which was strewed with misshapen rocks, scattered in rude confusion. With hearts nerved to a high resolve, the hapless pair awaited the arrival of their yelling pursuers. Conspicuous by his eagle plume, towering form and scowling brow, the daughter soon descried her inexorable sire, leaping from crag to crag below her. He paused abruptly when his fiery eye rested on the objects of his pursuit. Notching an arrow on the string of his tried and unerring bow, he raised his sinewy arms—but ere the missile was sent, Wun-nut-hay, *the Beautiful,* interposed her form between her father and his victim. In wild appealing tones she entreated her sire to spare the young chieftain, assuring him that they would leap together from the precipice rather than be separated. The stern old man, deaf to her supplication, and disregarding her menace, ordered his followers to seize the fugitive. Warrior after warrior darted up the rock, but on reaching the platform, at the moment when they were grasping to clutch the young brave, the lovers, locked in fond embrace, flung themselves

'From the steep rock, and perished.'

"The mangled bodies were buried in the bottom of the glen, beneath the shade of everlasting rocks; and two small hollows, resembling sunken graves, are to this day pointed out to the curious traveler, as the burial place of the lovers." It is a sweet, wild haunt, the sunbeams fall there with softened radiance, and a brook near by gives out a complaining murmur, as if mourning for the dead. [Footnote: Mr. Stone adds in a note— "This interesting legend was derived many years ago from a Seneca chief of some note, named Chequered Cap, and was communicated to me by W. H. C. Hosmer, Esq., of Avon. On the top of Genundewa the remains of an Indian orchard are visible, a few moss-grown and wind-bowed apple trees still linger, sad, but fitting emblems of the wasted race by whom they were planted."]

Let us return to the inquiry we were pursuing. Of the origin of the Iroquois confederacy, some traditionary accounts have been given, which represent the different tribes as dwelling for a time, in the separate locations assigned them, independent of each other. Here they increased in valor, skill and knowledge, suited to their forest home. At length becoming numerous, rival interests arose among them, which did not exist when they were small and feeble. They fell into contention, and wasted and destroyed each other. Each tribe fortified his own position, and dwelt in constant fear of being surprised and overcome by his neighboring foe.

At length one of their sachems, distinguished for his wisdom and address, proposed that they should cease from a strife, which was only destroying themselves, and unite their energies against the Alleghans, the Adirondacks, the Eries, and other ancient and warlike tribes, who were their superiors in their isolated and divided condition. Already weary of their unprofitable conflicts, the proposal was received with favor, and Ato-tar-ho, an Onandaga chieftain, unequalled in valor, and the fame of whose skill and daring was known among all the tribes, became the leading spirit of this confederacy, and by common consent was placed at its head. So fully did experience demonstrate the wisdom of this arrangement, that they used every means to strengthen the bands of their union, and by the most solemn engagements of fidelity to each other, they became the Ko-nos- hi-o-ni, or United people. [Footnote: Schoolcraft's Report.]

How long this confederacy had existed before their discovery by the whites, is unknown. There is a tradition which places it one age, or the length of a man's life, before the white people came to this country. [Footnote: Pyrlaus, a missionary at the ancient site of Dionderoga, or Fort Hunter, writing between 1742 and 1748, gives this as the best conjecture he could form, from information derived from the Mohawks. It is thought however that this time is too short, to account for the degree of development attained by the Iroquois, in their united capacity, at the time of their first discovery by the whites.]

The union of these several tribes was the means of securing their pre-eminence over the other Indians in this country. Their individual traits are thus very fittingly represented;—"in their firm physical type, and in their energy of character, and love of independence, no people among the aboriginal race have ever exceeded, if any has equalled the Iroquois." [Footnote: Schoolcraft.] They occupied a region surpassed by no other on the continent, for grandeur and beauty united, and inherited from this or some other source, a mental constitution of noble structure, which placed them in the fore-front of their race, and when united, no tribe on this continent could stand before them. This has served to render their history, a matter of earnest and interesting inquiry.

CHAPTER III

Name Red Jacket, how acquired—Indian name—Conferred name—Singular tradition—Red Jacket during the war of the Revolution—Neutrality of the Indians—Services sought by Great Britain—Sketch of Sir William Johnson— Position of Red Jacket—Taunt of cowardice—Testimony of Little Beard— Charge made by Brant—Red Jacket's indifference—Anecdote—Early love of eloquence—Interesting reminiscences.

The name Red Jacket, so familiar to the whites, was acquired during the war of the Revolution. He was distinguished at this time as well as afterward, for his fleetness on foot, his intelligence and activity. Having attracted the attention of a British officer by the vivacity of his manners, and the speedy execution of those errands with which he was intrusted, he received either in token of admiration, or for services rendered, or both, a beautifully ornamented jacket of a scarlet color.

This he took pride in wearing, and when worn out, he was presented with another, and continued to wear this peculiar dress until it became a mark of distinction, and gave him the name by which he was afterward best known. At a treaty held at Canandaigua in 1794, Captain Parrish, who was for many years agent of the United States for the Indians, presented him with another *red jacket* to perpetuate a name of which he was particularly fond. [Footnote: McKenney's Indian Biography Politely favored by Alfred B. Street, Esq., and assistant Mr. J. H. Hickox, of the State Library, Albany, N. Y.]

His original name was Oti-ti-ani, *always ready*. Sa-go-ye-wat-ha, the title conferred upon him at his election to the dignity of Sachem, has been rendered, "*The keeper awake, he keeps them awake, and the author, or cause of a wakeful spirit*." [Footnote: This latter translation was given to the author by the late Wm. Jones, a half-blood, son-in-law of Red Jacket and a chief of some note. This interpretation was given to some gentlemen from Buffalo who proposed to erect a monument at Red Jacket's grave. It was given in a full council of the chiefs of his tribe.]

The name is connected with a curious superstition among his people, and will best be understood, by an acquaintance with the circumstances under which it is used.

If during the still hours of night, an Indian's mind is taken up with thoughts that cause sleep to pass from him, preventing every effort of Morpheus to lock him in fond embrace, he ascribes it to a spirit, which he calls Sa-go-ye-wat-ha.

The impressions made are regarded as ominous of some important event, joyful or otherwise, according to the feelings awakened. If his thoughts are of a pleasing nature, he is led to anticipate the occurrence of some joyful event. If they are of a melancholy turn, he regards it as foreboding evil.

He may be led to dwell with interest on some absent friend; that friend he will expect to see the next day, or soon after. Yet should his thoughts be troubled or anxious, he would expect to hear soon of that friend's death, or that something evil had befallen him. [Footnote: Conversation with Wm. Jones, Seneca chief.]

Such was the spirit they called Sa-go-ye-wat-ha. He could arrest the current of their thought, bring before them visions of delight, or send upon them melancholy reflections, and fill their minds with anxiety and gloom.

This title conferred on Red Jacket, while it indicated the cause of his elevation, presented the highest compliment that could be paid to his powers of oratory. By the magic spell of his words, he could control their minds, make their hearts beat quick with emotions of joy, or send over them at will the deep pulsations of grief.

The incident referred to as giving rise to the name, Red Jacket, introduces him in connection with the war of the Revolution. As his conduct during this period has been the subject of frequent remark, severely criticised by some, and not very favorably viewed by others, justice to the orator's memory requires a brief statement of his reasons for the course he pursued.

While thoughts of this contest were pending, the colonists took measures to secure the favorable disposition of the Iroquois, and these efforts at the time were successful.

The general government advised them to remain neutral, during the anticipated conflict. This course met the approval of their most considerate sachems. For though inured to war, and apt to enter with avidity into the excitement of a conflict, their forces had been reduced by recent encounters with the Indians at the west, and south, and also with the French; and the few intervening years of peace served to convince them of its value, and caused them to receive with favor this proposition from our government.

At a council held with the Iroquois at German Flats, in June, 1776, by Gen. Schuyler, who had been appointed for this purpose, these assurances of neutrality were renewed.

Great Britain also was not indifferent about the course these Indian tribes would pursue. Wishing to prevent an alliance of the Indians with the colonists, willing to secure forces already on the ground, and with a view possibly, of striking terror into the minds of her rebellious subjects, her agents in this country spared no pains to enlist the sympathies of the Iroquois on her side.

In this they were but too successful. Through their agents, Britain had been in correspondence with these tribes for more than a hundred years, had supplied them with implements of war, articles of clothing, and with many of the comforts and conveniences of life. The Indians had learned to be dependent upon her, and they called her king their "*great father* over the water."Her agents spent their lives among them. Through them their communications were made to the crown, and they regarded them as essential to their happiness. Hence they exerted a very great influence over them.

This was especially true of Sir William Johnson, who died at Johnson Hall in the month of June, 1774.

Mr. Johnson was a native of Ireland, of a good family and fitted by nature and education, to adorn the walks of civilized life. He came to this country not far from 1738, as land agent of his uncle, Sir Peter Warren, an admiral in the English navy, who had acquired a considerable tract of land upon the Mohawk, in the present county of Montgomery.

Possessing a romantic disposition, he readily adapted himself to the rude customs that prevailed in the wilds of America.

The *Gentleman's Magazine* of London said of him in 1755,—"Besides his skill and experience as an officer, he is particularly happy in making himself beloved by all sorts of people, and can conform to all companies and to all conversations. He is very much of a gentleman in genteel company, but as the inhabitants next to him are mostly Dutch, he sits down with them and smokes his tobacco, drinks flip, and talks of improvements, bear and beaver skins. Being surrounded with Indians, he speaks several of their languages well, and has always some of them with him. He takes care of their wives, and old Indians, when they go out on parties; and even wears their dress. In short, by his honest dealings with them in trade, and his courage, which has often been successfully tried with them, he has so endeared himself to them, that they chose him as one of their chief sachems, or princes, and esteem him as their father."

Not far from the year 1755, while the French and English were at war, he was made general of the colonial militia, and by virtue of a leadership that had been created by the Iroquois, he was head warrior of all the Indian tribes, who favored the English.

The gifts of his sovereign, and the opportunity he had of purchasing Indian lands, were the means of his securing great wealth. The ease with which he secured land of the Indians is illustrated by an amusing occurrence between him and a noted chief, Hendrick. Soon after entering upon his duties as Superintendent of Indian Affairs in this country, he received from England some richly embroidered suits of clothes.

Hendrick, a Mohawk chief, was present, when the package containing them was opened, and could not refrain from expressing his admiration of them. He went away very thoughtful, but soon after returned and said to Sir William, that he had dreamed a dream.

"Ah! And what did you dream?" said Sir William.

"I dreamed," said Hendrick, "that you gave me one of those new suits of uniform."

Sir William could not refuse it, and one of the elegant suits was presented to Hendrick, who went away to show his gift to his countrymen and left Sir William to tell the joke to his friends. A while after the general met Hendrick and said—"Hendrick, *I have dreamed a dream*."

Whether the Sachem mistrusted he was now to be taken in his own net or not, is not certain, but he also inquired,—"And what did you dream?"

The general said he dreamed that Hendrick presented him with a certain piece of land which he described. It consisted of about five hundred acres, of the most valuable land in the Mohawk valley.

Hendrick replied,—"It is yours;" but, shaking his head, said, "Sir William I will never dream with you again." [Footnote: Drake's Book of the Indians.]

Sir William's large estate, the partiality of his countrymen, together with his military honors, and his great influence with the Indians, rendered him "as near a prince as anything the back-woods of America has witnessed." [Footnote: The expression of an English lady.—Turner.]

He built two spacious and convenient residences on the Mohawk river, known as Johnson Castle and Johnson Hall. The Hall was his summer residence. Here he lived something like a sovereign, kept an excellent table for strangers and officers, whom the course of duty led into these wilds, and by confiding entirely in the Indians, and treating them with truth and justice, never yielding to solicitations once refused, they were taught to repose in him the utmost confidence.

His personal popularity with the Indians, gave him an influence over them greater it is supposed, than any one of our own race has ever possessed. He was the first Englishman that contended successfully with French Indian diplomacy, as exercised by their governors, missionaries and traders. [Footnote: Turner's Phelps and Gorham Purchase.]

Had he lived until the war of the Revolution, it is supposed by some he might have remained neutral, and have kept the Indians from engaging in the conflict, though this is altogether uncertain. He lived to see the gathering of the storm that swept away most of his great possessions.

On the death of Sir William, his son John Johnson succeeded to his titles and estate. The office of General Superintendent of the Indians, fell into the hands of Col. Guy Johnson, a son-in-law, who appointed Col. Claus, another son-in-law, as his deputy.

Into their hands fell the property, and a large share of the influence over the Indians, possessed by Sir William Johnson. This influence was exerted in favor of Great Britain.

When the Indians heard of the uprising in Boston, and of the battle of Lexington, they were told, that these out-breaks were the acts of disobedient children, against the great king, who had been kind to them, as he had to the Six Nations. That their *"great father over the water,"* was rich in money and men; that the colonists were poor, and their numbers small, and that they could easily be brought into subjection.

At a council of the Iroquois convened at Oswego, by Sir John Johnson and other officers and friends of the crown, they were informed that the king desired them to assist him in subduing the rebels, who had taken up arms against him, and were about to rob him of a part of his great possessions.

But the chiefs one by one assured the British agents that they had the year before, in a council with General Schuyler, pledged themselves to neutrality, and could not without violating their promise, take up the hatchet.

But they were assured that the rebels justly merited all the punishment that white men and Indians could inflict;—that they would be richly rewarded for their services, and *that the king's rum was as plenty as the waters of Lake Ontario.*

This appeal to their appetites, already vitiated, together with the promise of large rewards, at length prevailed; and a treaty was concluded, in which the Indians pledged themselves to take up arms against the rebels, and continue in service during the war. They were then presented each with a suit of clothes, a brass kettle, a gun, a tomahawk, a scalping knife, a quantity of powder and lead, and a piece of gold. [Footnote: Life of Mary Jemison.]

The Senecas were among those who consented to join the royal standard. Of this action Red Jacket did not approve. He declared plainly and unhesitatingly to those who had determined to engage in the war,—"*This quarrel does not belong to us,—and it is test for us to take no part in it; we need not waste our blood to have it settled. If they fight us, we will fight them, but if they let us alone, we had better keep still.*" [Footnote: Testimony given to the author by Wm. Jones, Seneca chief, and confirmed by Col. Wm. Jones, son of the Indian interpreter, who affirms that prominent Indian chiefs had declared in his hearing that these were the sentiments of Red Jacket at this time.]

Red Jacket at this time was not far from twenty-six years old. His forensic abilities had not been called forth, and his influence weighed but little in comparison with that of older men. But it may be observed that his conduct ever after this, will be found consistent with the sentiments he entertained, and was free to express. Though young, his perceptions were keen, he had a deep and penetrating mind and saw at a glance that in this contest his people were doomed to suffer, to be ground between the upper and nether mill stone.

When, in the summer of 1777, his people received an invitation to join the forces that were preparing to march under the command of Col. St. Leger upon Fort Stanwix, being assured that they would not be required to endure the fatigues and dangers of the battle, but might "sit down quietly and smoke their pipes, and see the sport;" Red Jacket endeavored, but in vain, to prevent his people from going. He said to them, "*it's a cheat; the design is to deceive you, and if you go you will find that you have been deluded.*"

They threw back the taunt,—"*You are a coward, you have the mind of a woman, and are not fit to go to war.*"

Red Jacket though not at this time a chief, was a young man of acknowledged talent and influence, and having a right to express his opinion, did not hesitate to give it in favor of peace. His opinion was well known among his people. Little Beard has frequently been seen to bury his face in his blanket, and give vent to his tears, in view of the havoc made among the Senecas by the war, at the same time declaring,—"*Red Jacket was opposed to the war,* HE WAS ALWAYS IN FAVOR OF PEACE, *and how much better it had been, had we listened to his advice.*" [Footnote: Conversation of the author with Col. Jones.]

Red Jacket's prediction was too nearly verified. The Senecas suffered most severely in that campaign. They fell under the command of Thay-en-dan-e- gea or Brant, who went with a company of Tories, led by Col. Butler, to intercept General Herkimer, who was reported as coming to the relief of the garrison. At a certain point on the way, where they expected the general would pass, they formed an ambuscade, and though they selected their ground with wisdom, and acquitted themselves with great bravery, they were unable to stand before the invincible courage of the heroes of Oriskany.

The Senecas claim to have lost in that engagement thirty-three of their chiefs, and their feelings in view of it are said to have been sad in the extreme. [Footnote: "The mourning was excessive, and was expressed by the most doleful yells, shrieks and howlings and by inimitable gesticulations."—Mrs. Jemison's Narrative.]

The charge of cowardice applied by the young warriors to Red Jacket, upon their first starting out on this campaign, was one frequently made during the war. His views were at different times expressed in opposition to it, and his arguments as often repelled by the young braves, who could not endure his invectives. The reply was easily made, and hence in more frequent demand, than if it had imposed a greater tax upon their intellects. The epithet has often been applied to him since, and though his tastes did not lead him to seek the fame of a warrior, still it is believed he was not so devoid of courage, as has sometimes been represented.

His views of the war, were not those of a partisan, hence his conduct was often censured by those who had entered heartily into the contest.

Brant has charged him with being the occasion of trouble to him, in his efforts to arrest the march of Sullivan, and his army, into the Indian country. Particularly at Newtown, where considerable preparations had been made for defense. Says Col. Stone,—"Sa-go-ye-wat-ha was then twenty-nine years old, and though it does not appear that he had yet been created a chief, he nevertheless seems to have been already a man of influence. He was in the practice of holding private consultations with the young warriors, and some of the younger and less resolute chiefs, for the purpose of fomenting discontents, and persuading them to sue for what Brant considered, ignominious terms of peace.

"On one occasion as Brant has alleged, Red Jacket had so far succeeded in his treachery, as to induce some of the disaffected chiefs to send a runner into Sullivan's camp, to make known dissensions he himself had awakened, and invite a flag of truce, *with propositions of peace to the Indians*."

Though charged with acting criminally, it is here expressly asserted, *that it was to obtain peace*. Peace he most earnestly desired for his people, who were doomed to be wasted in a contest not their own.

Nor, in view of his feelings respecting the war, is it surprising he should have incurred the displeasure of Cornplanter, while endeavoring to bring his countrymen to make a stand against a portion of the invading army, on the beach of Canandaigua lake, where was an Indian village of some size. Not finding in Red Jacket an ardor for the undertaking which corresponded in any degree with his own, he turned to the young wife of the orator and exclaimed,—"*Leave that man, he is a coward; your children will disgrace you, they will all be cowards*." [Footnote: Col. Wm. Jones.]

The epithet thus applied occasioned uneasiness to none less than to the orator himself. Whenever he chose to notice it, he would make a good return for what he had received.—In a war of words, he was on his own chosen ground. He was a match for their greatest champion, and in cross-firing, it could easily be seen that his missiles were directed by one who was perfect master of the art. He could handle at will the most cutting sarcasm, and while maintaining a good natured, playful mood, deal his blows with such power and effect, as to make the victim of his irony resort to some other means of defense, than the tongue. It is said that frequently by his cool, good natured railery, he has caused the victim of his sport to turn upon and strike him. He would answer it by a hearty laugh, unless the blow was of such a nature as to demand of him a different reception. [Footnote: Wm. Jones, Seneca chief.] He seemed to be armed at every point, as with a coat of mail, against the arrows of his assailants. Their most powerful weapons would be turned aside by his presence of mind, and matchless skill, and leave him apparently unharmed.

A circumstance illustrating this point, once occurred between him and Little Billy, a chief of some note among the Senecas, who was frequently in the orator's company. This chief, with Red Jacket and one or two others, were once passing from their settlement on Canandaigua lake, to the old Seneca Castle, near the foot of Seneca lake. On their way they encountered a large grizzly bear. Little Billy and the others in the company, were frightened and began to run. Red Jacket who was distinguished as a hunter, and an excellent marksman, drew up his rifle, and brought the monster to the ground.

It so happened, on one occasion sometime afterward, that Little Billy was very pertinacious in calling Red Jacket a coward. The orator did not appear to notice him at first; but finding that he persisted in the charge, he turned to him and coolly and sarcastically said,—*"Well, if I am coward I never run unless it's for something bigger than a bear."* [Footnote: Conversation with Seneca chief, Wm. Jones.]

It is hardly necessary to add, that nothing more was heard from Little Billy concerning his cowardice on that day.

This charge of cowardice was owing in a great degree to the orator's position. He was not on the popular side. The majority of his people were against him. Had he acted in accordance with their wishes, it is a question whether anything would ever have been said about his deficiency in courage. And this supposition is strengthened by the fact, that at a subsequent period in his history, a little display of courage, when acting in accordance with the wishes of his people, gained for him a marked degree of approbation, and gave rise to the affirmation, *"the stain fixed upon his character, was thus wiped away by his good conduct in the field."* [Footnote: McKenney's Indian Biography.]

In opposing the wishes of his people, when bent on a war of which he did not approve, he gained the epithet of *coward*. With less intelligence, and less moral courage, he might have seconded the views of his nation, and been ranked a brave.

Hence, though we do not claim for Red Jacket the possession of qualities, adapted to make him conspicuous as a military chieftain, we are disposed to attribute to him the higher courage of acting in accordance with his own convictions of propriety and duty. "He was born an orator, and while morally brave, lacked the stolid insensibility to suffering and slaughter, which characterized the war-captains of his nation." [Footnote: Bryant's address.]

We readily concede that Red Jacket was fitted by nature to excel in councils of peace, rather than in enterprises of war; to gain victories in a conflict of mind with mind rather than in physical strife, on the field of battle.

And it may be questioned whether the qualities adapted to the highest achievements of oratory, would be congenial to the rough encounters of war. Especially when the mind is already preoccupied with inward thirstings after the glory of the rostrum; it will not be apt to sigh for the camp, or the noise and tinsel of mere military fame.

It is related of him that when a boy, he was present at a great council held on the Shenandoah. Many nations were there represented by their wise men and orators. The greatest among them was Logan, who had removed from the territory of his tribe to Shamokin. He was the son of Shikellemus, a celebrated chief of the Cayuga nation, who, before the Revolution was a warm friend of the whites.

On the occasion referred to, Red Jacket was so charmed with Logan's style, and manner of delivery, that he resolved to attain if possible the same high standard of eloquence; though he almost despaired of equalling his distinguished model.

On his return to Cunadesaga, near the Seneca lake, which was at that time his home, he sometimes incurred the displeasure and reproof of his mother, by long absence from her cabin, without any ostensible cause. When hard pressed for an answer, he informed his mother, that *"he had been playing Logan."*

"Thus in his mighty soul the fire of a generous emulation had been kindled, not to go out until his oratorical fame threw a refulgent glory on the declining fortunes of the once formidable Iroquois. In the deep and silent forest he practiced elocution, or to use his own expressive language, *played Logan,* until he caught the manner and tone of his great master. Unconsciously the forest orator, was an imitator of the eloquent Greek, who tuned his voice on the wild sea beach, to the thunders of the surge, and caught from nature's altar his loftiest inspiration.

"Not without previous preparation, and the severest discipline, did Red Jacket acquire his power of moving and melting his hearers. His graceful attitudes, significant gestures, perfect intonation, and impressive pauses, when the lifted finger, and flashing eye told more than utterance, were the result of sleepless toil; while his high acquirement was the product of stern habitual thought, study of man, and keen observation."

"He did not trust to the occasion alone for his finest periods, and noblest metaphors. In the armory of his capacious intellect the weapons of forensic warfare had been previously polished and stored away. Ever ready for the unfaltering tongue was the cutting rebuke, or apt illustration. By labor, persevering labor, he achieved his renown. By exercising his faculties in playing Logan when a boy, one of the highest standards of mortal eloquence, either in ancient or modern times, he has left a lesson to all ambitious aspirants, that there is no royal road to greatness; that the desired goal is only to be gained by scaling rugged cliffs, and treading painful paths." [Footnote: This statement, together with the remarks that follow, is presented almost entire, from a reminiscence of Red Jacket, given by Mr.

Turner in his Pioneer History of the Phelps and Gorham Purchase, a work that has rescued from oblivion, many interesting and valuable historical recollections.]

The habit thus acquired in the orator's youth, became characteristic of him, at a later period of his life. Previous to his making any great forensic effort, he could be seen walking in the woods alone, apparently in deep study. [Footnote: Col. Wm. Jones.]

CHAPTER IV

Early struggles—Red Jacket's opportunity for trial—Council at Fort Stanwix—Red Jacket's office of Sachem—Red Jacket's opposition to the proposed treaty—Excitement created by his speech—Allayed by Cornplanter—His influence in deciding the treaty—How it affected him.

How long and toilsome the way, ere the ambitious aspirant passes from the low grounds of obscurity, to the dazzling heights of fame! How many hours of anxious toil, through wearisome days and nights, protracted through months and years, are passed, before the arena even is entered, where the race commences in earnest! How many struggling emotions between hope and fear, encouragement and doubt, promise and despair, mark the experience, and clothe it with the sublimity and interest that belong to action in its highest forms!

Did this child of nature cherishing the bright dream from early life, never suffer from these contending emotions, ere he awoke finally to the consciousness of the reality, where he could exclaim, I am an orator, yes, I AM AN ORATOR!

This idea Red Jacket began now to cherish. He had practiced in his native wilds, the forest depths had echoed back those strains of eloquence, that had struggled for utterance in his impassioned bosom, and their force being expended here, served but to awaken a still stronger desire to try his powers, where he could have the answering sympathy of human hearts. His fame and greatness were yet to be achieved. With the inward consciousness of strength that would secure for him the eminence he desired, he awaited eagerly the opportunity for its exercise. This opportunity came.

When the storm of war had rolled by, the hour came for deliberation, and council. England and America had concluded peace, and the jurisdiction of the country of the Iroquois had been surrendered to the United States. Still no provision had been made by the crown for those tribes that had freely fought in her defence. They were left to make their own peace, or prosecute the war on their own account. Their attitude was yet hostile. No expedition of importance was undertaken, but the border men were constantly annoyed by Indians, who drove away their horses and cattle, and committed other acts of depredation. And the inhabitants of the frontier had suffered so severely from the Indian tribes during the war, that these acts served to awaken still deeper feelings of hostility toward them, and led some openly to recommend that the Indians be driven from their lands, and that these be forfeited to the State.

These councils were strenuously resisted by the general government. The humane and considerate Washington thought it wiser to try and conciliate them, and if possible win their confidence and esteem, claiming that their lands, when needed, could be obtained at a cheaper rate by negotiation and purchase, than by war and conquest.

This course, the excellence of which experience has fully demonstrated, was finally adopted, and in pursuance of this design, a general council of the Iroquois was convened at Fort Stanwix, in the fall of 1784. It was attended by Oliver Wolcott, Richard Butler, and Arthur Lee, who were appointed commissioners on the part of the United States. The different tribes of the Iroquois were represented, and Red Jacket was present, and took an active part in its deliberations. He had now been elected to the office of Sachem; at what time precisely, is not known, but probably not far from the close of the war of the Revolution.

The manner in which he gained this office has been ascribed by some to artifice as well as the force of his eloquence. Col. Stone says, that "aspiring to the rank of chief, he not only wrought upon the minds of his people, by the exertion of that faculty which was ever with them a high standard of merit, but he succeeded in availing himself of the superstitious constitution of his race, to effect his purpose. His first essay was to dream that he was, or should be a chief, and that the Great Spirit was angry that his nation had not advanced him to that dignity. This dream, with the necessary variations, he repeated until, fortunately for him, the small pox broke out among the Senecas. He then proclaimed the loathsome infliction a judgment sent by the Great Spirit, to punish them for their ingratitude to him. The consequence ultimately was, that by administering flattery to some, working upon the superstitious fears of others, and by awakening the admiration of all by his eloquence, he reached the goal of his ambition." [Footnote: Col. Stone's Life and Times of Thayendanegea and Life and Times of Red Jacket. This statement has been denied by some, who affirm that his eloquence was the sole cause of his elevation. If this representation came from Brant, it may be recollected that between Red Jacket and Brant there did not exist a very strong attachment, and statements made by one concerning the other, would not be likely to bear the coloring of a very warm friendship.]

However this may have been, it is certain this course was not necessary to establish Red Jacket's position among his people. The circumstances of their history created a necessity for his transcendent abilities, and the light of his genius, though it may have been obscured for a time, must eventually have shone forth, in its original beauty and splendor.

Red Jacket was now called upon to assist in the deliberations of his people, and from this time to the day of his death, we find him connected with, and bearing an important part in all of their public transactions.

The council at Fort Stanwix was the first occasion in which he appeared before the public. It was a meeting of no small moment. With an anxious heart the Indian left his home and wended his way, through his native forests, to the place where he was to meet in council, the chiefs of the thirteen fires. His own tribes had been wasted, by a long and bloody war. The nation they had so long clung to, and by whose artifice they had been led to engage in the strife, stood confessedly vanquished. A new power had arisen in the land, what bearing would it have on their future fortunes?

With the importance of this gathering none were more deeply impressed than Red Jacket.—Yonder he stands, alone;—his knit brow, and searching glance indicate a process of thought, which stirs deeply the emotions of the inner man.—Tread lightly, lest you disturb the silent evolutions of that airy battalion, that is wheeling into rank and file, thoughts that discharged in words, reach the mark and do execution.—Now he wears a look of indignation, which presently turns to one of proud defiance, as he contemplates the encroaching disposition of the white race.—Now you may detect an air of scorn, and his eye flashes fire, as he regards them at first a feeble colony, which might easily have been crushed by the strong arm of the Iroquois.—A feeling of deep concern directly overspreads his features, as he thinks of their advancing power, and of the prospect of their surpassing even the glory of his own ancestry.—A still deeper shade steals over him as he thinks of the waning fortunes of his people.— Presently his countenance is lighted up;—his feelings are all aglow,—a bright thought, has entered his mind.—He conceives the idea of the union of the entire race of red men, to resist the encroachments of the whites. —Are they not yet strong? And united, would they not yet be, a formidable power?

With anxious and matured thoughts, Red Jacket comes to this council gathering. Its bearing on his nation and race, he deeply scans, and treasures up those burning thoughts, with which he is to electrify, and set on fire the bosoms of his countrymen.

Of the proceedings of this council, little is known aside from the bare treaty itself. By this treaty perpetual peace and amity were agreed upon between the United States, and the Iroquois, and the latter ceded to the United States, all their lands lying west of a line commencing at the mouth of a creek four miles east of Niagara, at a place on Lake Ontario called Johnson's Landing; thence south, in a direction always four miles east of the portage, or carrying-path, between Lakes Erie and Ontario, to the mouth of Buffalo creek, on Lake Erie; thence due south to the north boundary of the state of Pennsylvania; thence west to the end of said boundary; thence south along the west boundary of the state of Pennsylvania to the Ohio river.

In consideration of this surrender to the United States of their claim to western lands, the Iroquois were to be secure in the peaceful possession of the lands they inhabited in the state of New York.

This treaty Red Jacket strenuously resisted. He regarded the proposed cession of lands as exorbitant and unjust, and summoned all the resources of his eloquence to defend his position. The course of his argument and the various means he took to enforce it, we have no means of adequately presenting. A few hints respecting it, and the testimony of those present as to the effect produced, is all we have to guide us in forming any estimate of its merits.

After giving a vivid representation of the encroachments already made upon them by the whites, and of the advances they were making in numbers and power, as well as extent of territory, he reminded his hearers of the ancient glory of the Iroquois, and contrasted it with their present wasted and feeble condition. They had been passing through a mighty convulsion, the hurricane had swept over their dwellings, their homes were laid waste, their country made desolate.

He directed them to the extensive dominion they had exercised. Their empire was wide, on the north, and east, and south, and west, there were none to stay their hand, or limit their power. A broad continent was open to them on every side, and their seats were large. But now they were met by a people to whom they had surrendered a large portion of their lands, and "they are driving us on toward the setting sun. They would shut us in, they would close up the path to our brethren at the west. We demand an open way."

They had no right, he affirmed, to part with their western lands. Their laws, their ancient usages forbade it. They ought never to decide a question so momentous as this, without giving all the parties a hearing, who have any interest in its decision. They should be present and join in their deliberations. Their brethren at the west had a right to be consulted in this matter.—It would be unworthy of the name, and exalted fame of the Iroquois, to decide the question without reference to them.— It was a question that affected deeply the interests of the entire race of red men on this continent. He declared finally that rather than yield to the exorbitant demands of the treaty, they should take up their arms, and prosecute the war on their own account.

Such is the scanty outline of a speech that made a wonderful impression on the minds of all his people who were present. During the progress of his speech, their emotions were wrought up to a pitch, that seemed to betoken a rising storm, and at times it seemed as though it needed but a spark to set on fire a flame that was ready to burst out with consuming force.

Those present, who did not understand the language of the orator, were deeply interested in his voice, his manner of elocution, and his perfect and inimitable action. They caught fire from his eye, and felt the inspiration, which was kindled in the minds of all who listened to him understandingly. When he sat down his work was accomplished. There was but one heart among his people. From this time on, he was the peerless orator of his nation.

A very interesting sketch of Red Jacket as an orator, refers, for the existence of the facts which form the basis of its statements, to a treaty held at Canandaigua in 1794. It has been copied by Drake, and published in almost every sketch of the orator's life. Mr. Stone questions its truthfulness on the ground that there is no notice of it in any notes of this council taken at the time, and because also there was evidently an absence of the peculiar circumstances, which the speech referred to, seems to demand. Still he introduces it under the supposition that if delivered there at all, it might have been during the excitement produced among the Indians, by the rejection from the council, by Col. Pickering, of one Johnson, a messenger from Brant, who had been invited to be present at that council. Yet this is by no means probable, as Red Jacket would have been far from rising into eloquence on an occasion, which from his known relations to the proud Mohawk, he would naturally view with satisfaction, instead of resentment. The more probable supposition is, that the writer caught up this as a traditionary statement, which, owing to the lapse of time and the uncertainty of memory, had been changed in one or two of its items, and receiving it as correct, penned it in good faith, as having transpired at that treaty. It is a correct presentation of some of the points in the orator's speech on this occasion, and is as follows: [Footnote: Mr. Stone justly supposes this speech might have been made at the treaty of Fort Stanwix in 1784.]

"... The witnesses of the scene will never forget the powers of native oratory. Two days had passed away in negotiation with the Indians for a cession of their lands. The contract was supposed to be nearly completed, when Red Jacket arose. With the grace and dignity of a Roman Senator, he drew his blanket around him, and with a piercing eye surveyed the multitude. All was hushed. Nothing interposed to break the silence, but the rustling of the leaves. After a long and solemn, but not unmeaning pause, he commenced in a low voice, and sententious style. Rising gradually with the subject, he depicted the primitive simplicity and happiness of his nation, and the wrongs they had sustained from the usurpations of white men, with such a bold and faithful pencil, that every auditor was soon roused to vengeance, or melted to tears. The effect was inexpressible. But ere the emotions of admiration and sympathy had subsided, the white men became alarmed. They were in the heart of an Indian country, surrounded by ten times their number, who were inflamed by a remembrance of their injuries, and excited to indignation by the eloquence of a favorite chief. Appalled and terrified, the white men cast a cheerless gaze on the hordes around them. A nod from the chiefs might be the onset of destruction. At this portentious moment, Farmer's Brother interposed. He replied not to his brother chief, but with a sagacity truly aboriginal, he caused the cessation of the council, introduced good cheer, commended the eloquence of Red Jacket, and before the meeting had reassembled, with the aid of other prudent chiefs, he had moderated the fury of his nation to a more salutary view of the question before them."

The commissioners replied, but without making much headway on account of the agitation and excitement, produced by the orator's speech; that by the common usages of war they might lay claim to a much larger extent of territory; that their demand was characterized by great moderation, and insisted on their yielding to the terms proposed.

There was little disposition among them to yield the point, yet the treaty was finally brought to a successful issue, by the influence of Cornplanter.

Cornplanter was a noble specimen of the Indian race. He had all the sagacity for which his people were distinguished, and was equally active, eloquent and brave. He was well qualified by his talents to engage in the legislative councils of his nation, and was unsurpassed by any, for prowess and daring in the bloody field of strife. No chief, Thayendanegea not excepted, had gained higher laurels for personal valor, and none commanded more fully the confidence and esteem of his nation. His people looked up to him as a tower of strength, and when he spake, his words fell upon them with the weight of great authority. Better acquainted than his junior associate with the details of war, and understanding likewise the wasted and feeble condition of his people, and having learned in the late conflict something of the power of the enemy they would have to encounter, he regarded the idea of their resistance as wholly impracticable, and advised a compliance with the terms of the treaty. Though he regretted the loss of any more territory, he wisely concluded it was better to lose a part, than to be deprived of all. And by throwing his influence decidedly in favor, he succeeded finally in quieting the minds of his people, and in persuading them to accede to the proposals made.

It is a matter of regret that so few traces are left, of Red Jacket's speech on this occasion. Yet had his speech been reported, we might have been as much at a loss as at present, to derive from it a just estimation of his talents. His speeches as reported are tame when compared with the effect produced.

The Indian was an unwritten language. The most distinguished orators of the Iroquois confederacy, matured their thoughts in solitude without the aid of the pen, and when uttered in the hearing of the people, they passed forever into oblivion, only as a striking passage may hare been retained in memory. And with them the want of a written language was thus in a measure compensated. They made an increased effort to treasure up their thoughts. Yet how much must necessarily have been lost! and how liable to waste away, that which remained.

Trusting to them how imperfect must have been a reported speech! And relying on those who transferred their speeches to a different language, we have little assurance of any thing better than mutilated transcripts of the original. Need we be surprised then, to find in Red Jacket's published speeches, a tameness unworthy of his fame? Red Jacket was esteemed by the men of his time as an orator, surpassingly eloquent.

In his speeches as reported, this does not appear. Hence, his reported speeches fail to do him justice, or the men of his time very much overrated his talents.

Taking the latter horn of the dilemma we impeach the judgment and good sense of those who have gone before us. Assuming the former, we present an admitted and proclaimed fact. His contemporaries, while they conceded to him the highest attributes and accomplishments of eloquence, unite in affirming that his reported speeches come far short of the original.

Captain Horatio Jones, a favorite interpreter, has frequently declared,—*"it is impossible to do Red Jacket justice."* The peculiar shade given to the idea, its beauty in its own native idiom, was often entirely lost in the transfer. In much the same way, Captain Jasper Parrish, of Canandaigua, has frequently been heard to speak, when referring to the forensic efforts of the orator.

And besides, those passages that were most deeply fraught with eloquence, were often lost entirely, from the fact that the way having been prepared by a recital of those details that are reported, the reporter himself has been carried away by the very flood that surrounded, uplifted, and carried away the mass of those who heard him speak. So that the only note that would be made, of a passage of considerable length, is given in one or two short sentences. [Footnote: Conversation of the author with Col. Wm. Jones.]

By the generality of the Iroquois, the terms of the treaty at Fort Stanwix were regarded as severe; and though the services of the renowned Cornplanter were engaged by the commissioners, in an effort to persuade the disaffected into a reconciliation with it, the attempt was but partially successful, and was made at the expense of his own high standing among his people. They were not easily reconciled, and were so much displeased with his conduct on this, and one or two subsequent occasions, that they even threatened his life. A circumstance he touchingly refers to in a speech addressed to General Washington.

"Father," said he, "we will not conceal from you that the great God and not man, has preserved Cornplanter, from the hands of his own nation. For they ask continually—where is the land which our children, and their children after them are to lie down upon? When the Sun goes down he opens his heart before God, and earlier than the sun appears upon the hills, he gives thanks for his protection during the night; for he feels that among men become desperate by their danger, it is God only that can preserve him."

CHAPTER V

Claim of the United States to Indian lands—Conflicting claims of different States—Difficulty settled—Attempt to acquire the land by a lease—Purchase by Phelps and Gorham—Further purchase by Robert Morris.

At the close of the war of the Revolution, the territory ceded by Great Britain to the United States, included large tracts of country occupied by the Indians. In ceding these lands, she ceded only the right claimed by herself, on the ground of original discovery, which was simply a priority of right to purchase of the original occupants of the soil. The Indians were allowed to dwell upon these lands, and were considered in a certain sense the owners, but were required in case of a sale, to dispose of them to the government. [Footnote: Kent's Commentary.]

As each State claimed to be sovereign in every interest not ceded to the general government, each State claimed the territory covered by its original charter. These charters, owing to great ignorance of geographical limits, created claims that conflicted with each other. From this source originated difficult questions about land titles and jurisdiction, between the States of Connecticut and Pennsylvania,—Massachusetts and New York. These difficulties which existed before, the greater question of the Revolutionary war suspended for a time, but when peace was concluded, they came up again for a consideration and settlement.

The way was in a measure prepared for this, by the relinquishment to the general government, on the part of New York in 1781, and of Massachusetts in 1785, of all their right to territory west of a meridian line drawn south, from the western end of Lake Ontario.

In the adjustment of these difficulties, Connecticut relinquished her claim to a tract of land on the Susquehanna in Pennsylvania, called the Gore, and acquired that part of the State of Ohio called New Connecticut, or Western Reserve. And Pennsylvania obtained a tract of land lying immediately beyond the western boundary of the State of New York, and north-east of her own, embracing the harbor of Presque Isle, on Lake Erie, familiarly known as the Triangle, thus giving her access to the waters of this Lake.

The question in controversy between the States of New York and Massachusetts was more serious, owing to the large amount of territory claimed by the latter in western New York. It was brought to an amicable settlement, by Massachusetts surrendering to New York the right of jurisdiction, over all the land west of the present eastern boundary of the State; and by New York giving to Massachusetts the pre-emptive right, or right of purchasing of the Indians, all of the lands lying west of a meridian line drawn through Seneca Lake, from a certain point on the northern boundary of Pennsylvania, reserving however, a strip of land one mile in width, along the eastern shore of the Niagara river. Thus New York, while she retained the sovereignty, lost the fee of about six millions of acres of land, in one of the finest regions of country in the new world. [Footnote: For a more full account, see "Turner's History of the Phelps and Gorham Purchase."]

While these difficulties were being adjusted, a magnificent speculation was in progress, which bid fair to meet the expectations of its earnest projectors. A company was organized, called the New York and Genesee Land Company, with a view to obtain the entire tract of Indian lands within the State. To evade the law forbidding the sale of these lands to any party not authorized by the State, it was proposed to obtain them by a lease, that should extend nine hundred and ninety-nine years. A lease extending so long, was regarded as equivalent to a sale.

With a view to further its designs another company, the Niagara Genesee Company, was also formed in Canada, of those who were most in correspondence with the Indians, and who would be influential in securing from them a decision in favor of their object.

These organizations, especially the New York Land Company, were large, and included men of wealth and prominence, both in New York and Canada. With such appliances as they were enabled to bring to bear upon the Indians, they secured, in November, 1787, a *lease for nine hundred and ninety- nine years*, of all the lands of the Iroquois in the State of New York, except some small reservations, and the privilege of hunting and fishing, for an annual rent of two thousand dollars, and a promised gift of twenty thousand dollars.

The formidable character of these associations created a just alarm, and measures were immediately undertaken to circumvent their influence. An act was passed by the Legislature of New York, in March, 1788, authorizing the governor to disregard all contracts made with the Indians, and not sanctioned by the State; and to cause those who had entered upon Indian lands under such contracts, to be driven off, and their houses destroyed.

The sheriff of the county was directed to dispossess intruders and burn their dwellings, and a military force was called out, that strictly enforced these orders.

Thus by the energetic action of Governor Clinton of New York, the designs of these organizations were overruled.

As early as 1784, the Legislature of New York had passed an act, appointing the governor, and a Board of Commissioners, the Superintendents of Indian affairs, and as there were other Indian lands within the State, not covered by the pre-emptive right of Massachusetts, these commissioners with the governor at their head, entered upon negotiations with a view of purchasing them, and securing a title to them for the State. [Footnote: The commissioners designated were: Abraham Cuyler, Peter Schuyler and Henry Glen, who associated with them Philip Schuyler, Robert Yates, Abraham Ten Broeck, A. Yates, Jr., P. W. Yates, John J. Beekman, Mathew Vischer, and Gen. Gansevoort.]

A council of the Iroquois was appointed for this purpose, at Fort Schuyler, on the first of September, 1788.

The Leasees disappointed and angered by the bold and decisive measures taken against them, exerted their influence to prevent the Indians from assembling. But by measures equally energetic in its favor, a representation of the different tribes was obtained, and a treaty was concluded on the 12th, in which was conveyed to the State the land of the Onondagas; some reservations excepted, in consideration of one thousand dollars, in hand paid and an annuity of five hundred dollars forever.

Then followed negotiations with the Oneidas. Speeches were interchanged, propositions made and rejected, until finally an agreement was made, and a deed of cession executed by the chiefs, conveying all their lands, excepting certain reservations, in consideration of two thousand dollars in money, two thousand dollars in clothing and other goods, one thousand dollars in provisions, five hundred dollars for the erection of a saw and grist mill on their reservation, and an annuity of six hundred dollars forever.

The commissioners next appointed a council to be held at Albany, December 15, 1788. Great difficulty was experienced in getting the Indians together, the Leasees it is said, "kept the Indians so continually intoxicated, it was impossible to do anything with them." [Footnote: Turner's History.]

It was not until the eleventh of the February following, that a sufficient number were brought together, to proceed with the negotiations; and on the twenty-fifth, the preliminaries having been settled, the Cayugas ceded to the State all of their lands, excepting a large reservation of one hundred square miles. It was in consideration of five hundred dollars in hand, sixteen hundred and twenty-eight dollars in June following, and an annuity of five hundred dollars forever.

Mr. Turner in alluding to these negotiations very properly observes, "it was only after a hard struggle of much perplexity and embarrassment, that the object was accomplished. For the honor of our country, it could be wished that all Indian negotiations and treaties, had been attended with as little wrong, had been conducted as fairly as were those under the auspices and general direction of George Clinton. No where has the veteran warrior and statesman left a better proof of his sterling integrity and ability, than is furnished by the records of these treaties. In no case did he allow the Indians to be deceived, but stated to them from time to time, with unwearied patience, the true conditions of the bargains they were consummating."

He says further, "the treaties for lands found the Six Nations in a miserable condition. They had warred on the side of a losing party; for long years the field and the chase had been neglected; they were suffering for food and raiment. Half-famished they flocked to the treaties and were fed and clothed. One item of expense charged in the accounts of the treaty at Albany in 1789, was for horses paid for, that the Indians had killed and eaten on their way down. For several years in addition to the amount of provisions distributed to them at the treaties, boatloads of corn were distributed among them by the State."

It does not appear that Red Jacket, Cornplanter, Brant, or other of the more noted chiefs among the Iroquois, were present to take a part in these negotiations. Hence exception was taken to these proceedings. When the time drew near for paying the first annuity, the Onondagas sent an agent to Governor Clinton, saying they had received four strings of wampum from the Senecas, forbidding them to go to Fort Stanwix to receive the money, and declaring also "that the governor of Quebec wanted their lands; that Sir John wanted them; Col. Butler wants the Cayugas' lands; and the commanding officer of Fort Niagara wants the Senecas' lands."

They were assured in reply that they might "make their minds easy," the governor would protect them; that the Leasees were the cause of their trouble.

The Cayugas also sent a message to the governor, saying they were "threatened with destruction, even total extermination. The voice comes from the west; *its sound is terrible, our brothers the Cayugas and Onondagas are to share the same fate.*"

The complaint was, they had sold their lands without consulting the *western tribes.*

The decided position of the Executive in giving them assurance of protection, was the means of dissipating their alarm.

Historical evidence renders it apparent, that at this early period, the design was entertained by those in Canada, whose control over the Indians was well nigh supreme, to gain through them possession of Western New York, and without compromising the government of Great Britain, sever it from the United States, connect it with the territory of the North-west, and hold it by Indian possession, in a sort of quasi allegiance, to the crown of England.

Their design with respect to Western New York was defeated by the energetic measures of its chief executive, but further on we will see they did not relinquish the idea of holding from the United States, the territory of the North-west.

Next in the race of competition for the broad and fertile lands of the Genesee, appear the names of Oliver Phelps and Nathaniel Gorham. They were the acknowledged representatives of a considerable body of men, who were ambitious of securing an interest in what was regarded as the most desirable region in this country.

From the advent of Gen. Sullivan's army into the Indian country in 1779, their route being through the very finest portion of Western New York, and at a season of the year when vegetation was in its highest perfection; the beauty and fertility of these lands became the theme of praise, on the part of every soldier that beheld them. Their fame was thus carried to almost every village and hamlet in Pennsylvania and New England. Hence great eagerness was manifested in regard to the title, and settlement of these lands.

The company of which Messrs. Phelps and Gorham were the leading spirits, having purchased the pre-emptive right of Massachusetts, in the spring of 1788, Mr. Phelps went on to the ground, and was successful in convening a council of the Indians for the sale of their lands, at Buffalo creek, during the month of July of the same year. [Footnote: His success in obtaining this council, and securing a sale, was owing in a large degree, to his policy in paying court to the powerful faction of the Leasees residing in Canada, and giving them an interest in the purchase.]

The Indians at this treaty strenuously resisted the sale of any of their land west of the Genesee river; yet with a view of furnishing *"a piece of ground for a mill yard"* at the Genesee Falls, were finally persuaded to give their assent to a boundary line, that included a tract twelve miles square, west of that river. The eastern boundary of the lands sold, was the Massachusetts pre-emptive line; the western, was a line "beginning in the northern line of Pennsylvania, due south of the corner or point of land made by the confluence of the Genesee river, and the Canaseraga creek, thence north on said meridian line to the corner or point, at the confluence aforesaid; thence northwardly along the waters of the Genesee river, to a point two miles north of Canawangus village, thence running due west 12 miles; thence running northwardly so as to be twelve miles distant from the western bounds of said river, to the shores of Lake Ontario." The lands thus ceded, are what has been called *"The Phelps and Gorham Purchase."* It contained by estimation two million and six hundred thousand acres, for which they agreed to pay the Indians five thousand dollars, and an annuity of five hundred dollars forever.

Robert Morris, the distinguished financier of the Revolution, afterward became owner of the greater part of this purchase, as well as of the pre-emptive right of Massachusetts to the remaining part of Western New York. Through his agent in London, Wm. Temple Franklin, grandson of Doctor Franklin, these lands were again sold to an association of gentlemen, consisting of Sir William Pultney, John Hornby, and Patrick Colquhoun, and the farther settlement of this region, auspiciously commenced under its original proprietors, was conducted principally under their administration.

An intelligent and enterprising young Scotchman, Charles Williamson, who had previously devoted his time while detained as a prisoner in this country, during the war of the Revolution, to investigations respecting its geographical resources and limits, and who from his disposition and business capacity, was well qualified for the station, was appointed their agent, and emigrating hither with his family, and two other young Scotchmen as his assistants, John Johnstone, and Charles Cameron, he became identified with the early history and progress of the extensive and important part of the Indian territory, that as we have seen, had just been opened, and was inviting a new race, to take possession of its virgin soil.

CHAPTER VI

Union of the Western Indian Tribes contemplated—Hostile influence of the agents of Great Britain in Canada—Ambitious project of Thayendanegea or Brant—Council at Tioga Point—Indian Ceremonies—Visit of Cornplanter and others at the seat of government—Kindly feeling of Washington—Fresh occasion of trouble.

When Red Jacket, at the treaty of Fort Stanwix, in 1784, projected the bold idea of the union of all the Indian tribes on the continent, to resist the aggressions of the whites, he may not have thought it would soon come near having a practical fulfillment. This thought grew out of the circumstances and necessities of the times, and was the natural forecast of a great mind. His words sank deep into the hearts of his people,—they were carried beyond the bounds of that council-fire,—they went gliding along with the light canoe that plied the Lakes,—and were wafted onward by the waters of the Ohio and Mississippi. Several causes contributed to give direction and force to this movement.

Prominent among them was the fact, that the treaty of peace with Great Britain in 1783, though it put an end to the war, did not secure friendly relations between the two countries. Hostile feelings had been engendered and were still cherished, particularly by those who had taken refuge in Canada, in the early part of the Revolutionary struggle. Some of them were very active in stirring up Indian hostilities among the tribes at the west.

But prominent above all others were the exertions of Thayendanegea, or Brant, the famous war-chief, from whose leadership the inhabitants of our frontier settlements had suffered so severely, during the war of the Revolution. Very soon after the treaty at Fort Stanwix in 1784, from the dissatisfaction growing out of that treaty, and other indications among the Indians, he began to entertain the ambitious project of forming a grand Indian confederacy, of which he would be chief, embracing not only the Iroquois, but all of the Indian nations of the great North-west. He had given the entire summer of 1785, to the business of visiting these nations, and holding councils among them, with a view to the furtherance of this object. [Footnote: See Stone's Life and Times of Brant, Vol. 2, p. 248.]

He visited England at the close of this year, "ostensibly for the purpose of adjusting the claims of the loyal Mohawks upon the crown, for indemnification of their losses and sacrifices in the contest, from which they had recently emerged." [Footnote: See Stone's Life and Times of Brant, Vol. 2, p. 248.]

... "Coupled with the special business of the Indian claims, was the design of *sounding the British government, touching the degree of countenance or the amount of assistance which he might expect from that quarter, in the event of a general Indian war against the United States*." [Footnote: Ibid.]

His arrival at Salisbury was thus noted in a letter from that place, dated December 12, 1785, and published in London. "Monday last, Colonel Joseph Brant, the celebrated King of the Mohawks, arrived in this city from America, and after dining with Colonel De Peister, at the head-quarters here, proceeded on his journey to London. This extraordinary personage is said to have presided at the late Grand Congress of Confederate chiefs, of the Indian nations in America, and to be by them appointed to the conduct and chief command in the war, *which they now meditate against the United States of America*. He took his departure for England immediately as that assembly broke up; and it is conjectured that his embassy to the British Court is of great importance." [Footnote: Life of Brant, Vol. 2, p. 249.]

No public, decisive answer, for obvious reasons, was given to this application for countenance and aid in the contemplated war, for this part of the errand of the Mohawk chief, was "*unknown to the public at that day*." [Footnote: Life of Brant, Vol. 2, p. 249.]

Captain Brant on his return to America in 1786, entered once more upon the work of combining the Indian forces, and assembled a grand confederate council, which was held at Huron village, near the mouth of Detroit River. [Footnote: It was attended by the Six Nations, the Hurons, Ottawas, Miamis, Shawanese, Chippewas, Cherokees, Delawares, Pottowattamies, and Wabash, confederates.]

An address to the Congress of the United States was agreed upon at this council, pacific in its tone, provided no encroachments were made upon their lands west of the Ohio river. This was their ultimatum previous to the war, in which they were afterwards united.

At the treaty of peace between Great Britain and the United States in 1783, it was stipulated that the military posts south of the great lakes should be surrendered. This surrender was refused, on the plea that the United States had not fulfilled an agreement on her part, to see the just claims, due the subjects of Great Britain, cancelled.

From certain correspondence at this time it appears that there were other reasons also, for the witholding of these forts. Their surrender was earnestly desired on the part of the United States, as it was well understood, they gave encouragement to the hostile combinations, that at this time were going on.

In a letter to Captain Brant by Sir John Johnson dated Quebec, March 22d, 1787, he says, "Do not suffer an idea to hold a place in your mind, that it will be for your interest to sit still and see the Americans attempt the posts. [Footnote: Oswegatchie, Oswego, Niagara, Detroit and Mackinaw.] *It is for your sakes chiefly, if not entirely, that we hold them.* If you become indifferent about them, they may perhaps be given up; what security would you then have? You would be at the mercy of a people whose blood calls aloud for revenge; whereas, by supporting them, you encourage us to hold them, and encourage new settlements, already considerable, and every day increasing by numbers coming in, who find they can't live in the States. Many thousands are preparing to come in. This increase of his Majesty's subjects will serve as a protection to you, should the subjects of the States, by endeavoring to make further encroachments on you, disturb your quiet." [Footnote: Stone's Life and Times of Brant.]

Another letter soon after, by Major Mathews seems to confirm the above statements. "His Lordship [Footnote: Lord Dorchester, Governor General of Canada, formerly Sir Guy Carlton.] wishes them (the Indians), to act as is best for their interest; he cannot begin a war with the Americans, because some of their people encroach and make depredations upon parts of the Indian country; but they must see it is his Lordship's *intention to defend the posts*; and that while these are preserved, the Indians must find great security therefrom, and consequently the Americans greater difficulty in taking their lands; but should they once become masters of the posts, they will surround the Indians, and accomplish their purpose with little trouble." [Footnote: Life of Brant, Vol. 2, p. 271.]

Thus it is seen that those at the head of British affairs in Canada, while they studiously avoided coming into open collision with the United States, were viewing with satisfaction the gathering war-cloud, and were lending their influence to extend and intensify its threatening character.

The only course left for the United States was to prepare for the conflict; and while forces were being summoned to take the field, they were preceded by efforts of a pacific character.

A treaty was held with the Six Nations at Fort Harmar, on the Muskingum, in January, 1789, by Gen. St. Clair, in behalf of the United States, with a view to renew and confirm all the engagements, made at the treaty of Fort Stanwix in 1784. Goods amounting to three thousand dollars were distributed among the Indians, after the satisfactory conclusion and signing of the treaty. [Footnote: Indian treaties.]

At the same time a treaty was concluded with the Wyandot, Delaware, Ottawa, Chippewa, Pottowattamie and Sac nations, and goods distributed among them amounting to six thousand dollars, for a relinquishment of their claim to western lands.

These negotiations were doubtless attended with a beneficial influence, but they could not arrest the tide of warlike feeling that had been created. Hostilities were continued throughout the long line of our frontier settlements, and two of the Senecas having been killed by some bordermen of Pennsylvania, a great excitement was awakened among them.

Our government, anxious to remove the new occasion of disaffection, immediately disavowed the act, sought to bring the perpetrators of the crime to justice, and invited a friendly conference of the Iroquois at Tioga Point.

This council was convened on the sixteenth and remained in session until the twenty-third of November, 1790.

The chiefs in attendance at this council, and who took an active part in its deliberations, were Fish Carrier, Farmer's Brother, Hendrick, Little Billy and Red Jacket.

Colonel Pickering, as commissioner on the part of the United States, was present.

Red Jacket, their principal speaker, portrayed in a vivid and strong light, the sorrow they experienced, the injustice they had suffered, and the unpleasant feelings aroused among them. A large number of Indians were present, and were powerfully moved, and deeply affected by his speech.

Colonel Pickering, on the other hand, gave a very clear view of the facts in the case, showing conclusively the innocence of the government in the murder committed, and after a time succeeded in allaying the excitement, drying up their tears, and wiping out the blood that had been shed.

This council was enlivened by good cheer, and the observance of ceremonies common among the Indians.

Thomas Morris, who was present, was at this time adopted into one of their tribes. His father, Robert Morris of Philadelphia, having purchased of Massachusetts, in 1790, the pre-emptive right to that part of Western New York, not sold to Phelps and Gorham, sent his son, as preparatory to the negotiations he desired to make with the Indians, and for the general management of his business connected with the undertaking, to reside in

Canandaigua. While here he was diligent in cultivating an acquaintance with the principal chiefs of the Iroquois confederacy, who resided in that region. In this he was successful, and soon became a general favorite among them. He was in attendance with Colonel Pickering at Tioga Point, where the Indians determined to adopt him into the Seneca nation, and Red Jacket bestowed upon him the name himself had borne, previous to his elevation to the dignity of Sachem; O-ti-ti-ani, "*Always Ready*." It is beautifully described by Colonel Stone, and is given in his language.

"The occasion of which they availed themselves to perform the cermony of conferring upon young Morris his new name, was a religious observance, when the whole sixteen hundred Indians present at the treaty, united in an offering to the moon, then being at her full. It was a clear night, and the moon shone with uncommon brilliancy. The host of Indians, and their neophite, were all seated upon the ground in an extended circle, on one side of which a large fire was kept burning. The aged Cayuga chieftain, Fish Carrier, who was held in exalted veneration for his wisdom, and who had been greatly distinguished for his bravery from his youth up, officiated as the high priest of the occasion;—making a long speech to the luminary, occasionally throwing tobacco into the fire, as incense. On the conclusion of the address, the whole company prostrated themselves upon the bosom of their parent earth, and a grunting sound of approbation was uttered from mouth to mouth, around the entire circle.

"At a short distance from the fire a post had been planted in the earth, intended to represent the stake of torture, to which captives are bound for execution. After the ceremonies in favor of Madam Luna had been ended, they commenced a war-dance around the post, and the spectacle must have been as picturesque as it was animating and wild. The young braves engaged in the dance were naked, excepting a breech-cloth about their loins. They were painted frightfully, their backs being chalked white, with irregular streaks of red, denoting the streaming of blood. Frequently would they cease from dancing, while one of their number ran to the fire, snatching thence a blazing stick, placed there for that purpose, which he would thrust at the post, as though inflicting torture upon a prisoner. In the course of the dance they sung their songs, and made the forests ring with their wild screams and shouts, as they boasted of their deeds of war, and told the number of scalps they had respectively taken, or which had been taken by their nation. During the dance those engaged in it, as did others also, partook freely of unmixed rum, and by consequence of the natural excitement of the occasion, and the artificial excitement of the liquor the festival had well nigh turned out a tragedy. It happened that among the dancers was an Oneida warrior, who in striking the post, boasted of the number of scalps taken by his nation during the war of the Revolution. Now the Oneidas, it will be recollected, had sustained the cause of the colonies in that contest, while the rest of the Iroquois confederacy, had espoused that of the crown. The boasting of the Oneida warrior therefore, was like striking a spark into a keg of powder. The ire of the Senecas was kindled in an instant, and they in turn boasted of the number of scalps taken from the Oneidas in that contest. They moreover taunted the Oneidas as cowards. Quick as lightning the hands of the latter were upon their weapons, and in turn the knives and tomahawks of the Senecas began to glitter in the moon-beams, as they were hastily drawn forth. For an instant it was a scene of anxious, almost breathless suspense, a death- struggle seeming inevitable, when the storm was hushed by the interposition of Old Fish Carrier, who rushed forward, and striking the post with violence, exclaimed '*You are all a parcel of boys. When you have attained my age, and performed the warlike deeds that I have performed, you may boast of what you have done; but not till then.*'

"Saying which he threw down the post, put an end to the dance, and caused the assembly to retire. This scene in its reality must have been one of absorbing and peculiar interest. An assembly of nearly two thousand inhabitants of the forest, grotesquely clad in skins and strouds, with shining ornaments of silver, and their coarse raven hair falling over their shoulders, and playing wildly in the wind as it swept past, sighing mournfully among the giant branches of the trees above, such a group gathered in a broad circle of an opening in the wilderness, the starry canopy of heaven glittering above them, the moon casting her silver mantle around their dusky forms, and a large fire blazing in the midst of them, before which they were working their spells, and performing their savage rites, must have presented a spectacle of long and vivid remembrance." [Footnote: Stone's Life and Times of Sa-go-ye-wat-ha.]

This meeting conducted with evident good feeling, served much to allay the excitement and anger of the Senecas, and other tribes there represented, but the question concerning their lands, was still agitated and created dissatisfaction.

With a view to obtain some concession in their favor, Cornplanter, Half Town and Big Tree visited Philadelphia, which was at that time the seat of the general government, very soon after the council at Tioga Point. They were especially anxious to obtain the restoration of a portion of land south of Lake Erie, and bordering upon Pennsylvania, which was occupied by Half Town and his clan. They represented it as the land on which Half Town and all his people live, with other chiefs who always have been, and still are dissatisfied with the treaty at Fort Stanwix. "They grew out of this land, and their fathers grew out of it, and they cannot be persuaded to part with it. *We therefore entreat you to restore to us this little piece."*

This appeal, so simple and touching, was responded to by President Washington with great kindness. He reminded them that the treaty at Fort Stanwix had been fully confirmed at Fort Harmar in 1789, that it was not within his province to annul the provisions of a treaty, especially one that had been concluded before his administration commenced, yet he assured them that Half Town and his people, should not be disturbed, in the peaceful occupancy of the land in question.

From the friendly manner in which they were received and treated by the President, and the generous gifts bestowed, they returned home feeling satisfied that the ruler of the thirteen fires would do them no injustice, and they were hence better reconciled to the people he governed. Before leaving, however, they were engaged to go in company with Colonel Proctor, of the Indian Department, on an embassy of peace to the hostile tribes at the West, which was undertaken in the following spring.

On reaching their own country it was found that another outrage had been committed by a party of border-men, upon the Senecas at Beaver Creek, in the neighborhood of Pittsburg, in which three men and one woman were killed.

Cornplanter immediately sent runners with a dispatch to the government, informing them of the event, and with the earnest inquiry, "Our father, and ruler over all mankind, *speak now and tell us, did you order those men to be killed*?"

The secretary of war utterly disclaimed and denounced the transaction, promised them restitution, and that the offenders should be brought to justice. These times were so fruitful in difficulties, that ere one was healed another was created; yet our government by wise and prompt measures were after this successful, in securing peace with all of the Iroquois family within its borders.

CHAPTER VII

Expedition under Gen. Harmar—Its failure—High hopes of the Indians—Col. Proctor's visit to the Indians at Buffalo Creek—Red Jacket's speech—Indian deputation refused—Interference of the matrons—Council at Painted Post—Chiefs invited to Philadelphia.

The efforts of our government to secure peace with the Indians, were but partially successful. As our settlements extended westward in Pennsylvania, and along the Ohio and Kentucky borders, Indian hostilities and depredations continued to multiply. From the year 1783 when peace was concluded with Great Britain, until October, 1790, when the United States commenced offensive operations against them, the Indians killed, wounded and took prisoners on the Ohio and the frontiers, about fifteen hundred men, women and children; besides taking away two thousand horses, and a large amount of other property.

The Shawanese, Miamis and Wabash Indians were chiefly concerned in these bloody transactions; and our government finding protection for her citizens could not be secured by pacific means, resolved to proceed with vigorous offensive measures.

General Harmar, a veteran of the Revolution, with a force of fourteen hundred and fifty men, three hundred and twenty from the regular army, and the balance made up of recruits from Kentucky and Pennsylvania, advanced toward the Indian country.

The expedition left Fort Washington, the present site of Cincinnati, on the 30th of September, 1790.

The Indians, who kept watch of his movements, burned before his coming, their principal village and retired. Seizing a favorable opportunity, they fell suddenly upon a detachment of the main army commanded by Colonel Harding, consisting of two hundred and ten men, thirty of whom were regulars.

At the first onset the militia, the main part of the force, fled. The regulars stood their ground bravely for a time, but at a fearful odds; seven only escaped.

Colonel Harding, desirous of retrieving the disgrace, the next day with three hundred militia and sixty regulars, gave battle to the Indians. They fought near the junction of St. Joseph and St. Mary rivers, and the struggle, though severe and bloody, ended with the defeat of the Americans.

Success elated and emboldened the Indians, and rendered hopeless the negotiations for peace. Nevertheless the mission of Colonel Proctor, with a deputation from the Iroquois was not given up, and when spring opened he repaired to their country, to be joined by Cornplanter, Red Jacket and others, and proceed on his visit to the hostile tribes of the West.

Anticipating his arrival, a council of the Iroquois had been called at Buffalo Creek. Already messengers had been sent, earnestly soliciting them to join the warriors that were rising everywhere, from toward the setting sun. They had defeated the Americans, and nothing was wanting but the united action of all the Indian tribes, to secure the broad lands of the North-west, where they could spread their blankets in peace, and dwell securely forever. The Senecas, particularly, were urged to join in a war, that opened so many hopeful and glorious anticipations. The distinguished warrior Brant was very solicitous on this point, and being encouraged by those at the head of British affairs in Canada, was sanguine of ultimate success.

Colonel Proctor, accompanied by Cornplanter, arrived at the council fire kindled at Buffalo Creek, on the 27th of April, 1791.

Among the Indian chiefs present were Young King, Farmer's Brother and Red Jacket. The latter had now an acknowledged pre-eminence among his people, and took a leading part in the deliberations of this council. It was opened by a speech from Red Jacket, as follows:

"Brother: Listen. As is our custom we now address you, and we speak to you as to a brother that has been long absent. We all address you, and our chief warrior, Cornplanter; and we thank the Great Spirit for his and your safe arrival, coming as you do hand in hand from Honandaganius [Footnote: Name given to General Washington.] on important business.

"You have traveled long with tearful eyes, from the roughness of the way, and the inclemency of the season. Besides the difficulties between the bad Indians and our brothers the white people, everything has been conspiring to prevent your coming, thwart your business, and cause you to lose your way. The great waters might have prevented your coming; the wars might have stopped you; sickness and death might have overtaken you, for we know not what is to happen till it comes upon us. Therefore we thank the Great Spirit, who has preserved you from dangers, that would have prevented our hearing the good news you have come to bring us. And when filled with good news, how is it possible that disasters should befall you on the way?

"Wipe therefore from your eyes, the tears that have been occasioned by the dangers through which you have come. We now place you upon a seat where you can sit erect, a seat where you will be secure from the fear of your enemies, where you can look around upon all as your friends and brothers in peace.

"You have come with your heart and lips firmly closed, lest you should lose anything you had to say. With a brotherly hand we now open your hearts, and we remove the seal from your lips, that you may open them and speak freely without obstruction. Your ears too have been closed, that they might hear nothing until saluted by our voice. Open your ears to hear our counsels when we shall have had messages from you.

"We present therefore the compliments of the chiefs and head men of Buffalo Creek, to you and to our great warrior, the Cornplanter, hoping that you may each proceed safely with your business."

To this Cornplanter replied briefly, in behalf of himself and Col. Procter, reciprocating the kindness manifested, in the welcome that had been given them.

After which Col. Proctor explained fully the object of their coming, which was to obtain from them a deputation of peace, to visit with him the hostile Indian tribes at the West; and assured them of the liberal views, and friendly feelings of the chief of the thirteen fires toward them.

Several days were thereupon consumed in devising expedients and raising objections, which terminated finally in the declaration that nothing could be done without consulting their *British friends at Fort Niagara*. They desired the colonel to go with them there. His business not being with the British, but with them, he declined going. They then insisted upon having one of the officers of the fort to sit with them in council.

This being allowed, Col. Butler afterward appeared among them, and after a little private consultation with him, they seemed to be utterly averse to sending the proposed delegation.

Captain Brant, just before starting on a visit to western tribes, had been holding a consultation with these chiefs, and had no doubt been influential in causing them to be averse to joining this embassy.

Col. Proctor, finding further negotiation hopeless, declared his purpose to return, and expressed his regret in having to carry back an unfavorable report to the government, on whose kind and pacific errand he had been sent forth.

This announcement made a deep impression on their minds, and immediately a change took place in their proceedings, which revealed a peculiar feature of Indian diplomacy.

The women, who had been carefully watching the proceedings of this council, began to express their unwillingness to send to General Washington an unfavorable reply. To them was conceded the right, in things pertaining to the safety of their homes, of reversing, if they thought proper, the decision of the men. They did so on this occasion, and employed Red Jacket to present their views on the following day.

It was decided by them, in view of the threatening aspect of affairs, that Cornplanter, their most experienced warrior, should not leave them; but that a sufficient deputation, for which they had obtained volunteers, should accompany Col. Proctor, at the same time advising him of the danger, admonishing him to proceed with caution; *"to reach his neck over the land, and take in all the light he could, that would show him his danger."*

The journey being regarded as too hazardous by land, and the Indians unwilling to perform it with their canoes, the case was decided by the British officers, who refused them a vessel for the undertaking.

So great was the excitement among the Indians at this time, that before the result of Col. Proctor's mission was generally known, another council of the Iroquois was invited to meet at Painted Post, and was held during the month of June following. The British officers at Niagara, and runners from the western tribes, exerted their influence to prevent the Iroquois from coming into alliances of peace with the United States. But through the exertions of Col. Proctor, assisted by Cornplanter and the elderly matrons, the minds of the leading chiefs were turned from the proposed western alliance to Colonel Pickering and the treaty ground at Painted Post.

Red Jacket, together with other leading chiefs was present, and took an active part in the deliberations of this council. It was well attended by the Indians, as also by several American gentlemen, and a number of speeches were interchanged, whose general drift was in the direction of peace.

The result of this gathering was satisfactory to all parties. It served to bind more closely the friendship of the leading chiefs to the United States, and it served also to interest the minds of the young warriors, who had else from a love of adventure followed the war path, with the tribes at the West.

At the close of this council, a large entertainment was prepared purely after the civilized style, and when it was about concluded, Colonel Pickering took occasion to place before them the blessings and advantages of a cultivated state of society; and the happy influences that would arise from the introduction among them of the arts of civilized life. He assured them of the kind interest felt by General Washington and others in their welfare, and promised to aid them in any efforts they were disposed to make, for the advancement of their people. Presents were then liberally distributed among them, and they were invited at a convenient time to visit General Washington, and confer with him more fully on the subject.

The Indians were pleased with these suggestions, and promised to accept of the proffered invitation. Thus happily closed this council, gathered amid distracting influences, the Indians returning home better satisfied with their friendly attitude toward the government, and their feelings in striking contrast with those of their brethren at the West.

CHAPTER VIII

Expedition to the Indian Country—Washington's charge to Gen. St. Clair—Approach to the Indian villages—Sudden surprise—Disasterous battle—Indian victory—Retreat of the Americans—Boldness of the Indians—Friendly Indian deputation—Welcome of the governor of Pennsylvania—Red Jacket's reply—Address of President Washington—Reply of Red Jacket—Cause of Indian hostilities.

Indian hostilities still continued to destroy the peace and safety of our frontier settlements. And Congress with a view to provide relief, resolved to increase our military force, and place in the hands of the Executive, more ample means for their defense. A new expedition was therefore projected. General St. Clair, governor of the territory west of the Ohio, was appointed commander-in-chief of the forces to be employed.

President Washington had been deeply pained by the disasters of General Harmar's expedition to the Wabash, resulting from Indian ambushes. In taking leave of his old military comrade, St. Clair, he wished him success and honor; at the same time to put him on his guard, said,—"You have your instructions from the secretary of war. I had a strict eye to them, and will add but one word—Beware of a surprise! You know how the Indians fight. I repeat it—*Beware of a surprise*!" With these warning words sounding in his ear, St. Clair departed. [Footnote: Irving's life of Washington.]

On the seventh of September, 1791, General St. Clair set out for the Indian country. The American banner was unfurled and waved proudly over two thousand of her soldiers, as with sanguine hopes and bright anticipations, they took up their line of march for the Miami, designing to destroy the Indian villages on that river, expel the savages from the region, and by establishing a line of posts to the Ohio river, prevent the Indians from returning to a point, where they had been the occasion of great mischief. On their way they constructed two forts, Hamilton and Jefferson, and advanced but slowly, having to open for themselves a way through the forest. Too many of those composing this little army were deficient in soldier-like qualities. They had been recruited from the off- scourings of large towns and cities, enervated by idleness, debauchery, and every species of vice, which unfitted them for the arduous service of Indian warfare.

Hence insubordination, and frequent desertion, were among the difficulties encountered.

Not until the third of November, did they come near the Indian villages on the Miami. On the evening of that day they selected a position on the bank of a creek, which favored their purpose, and bivouacked for the night. Their number, from desertion, and those left to garrison the forts, amounted to but fourteen hundred. The place of their encampment was surrounded by close woods, dense thickets, and the trunks of fallen trees, affording a fine cover for stealthy Indian warfare.

It was the intention of St. Clair to throw up a slight work on the following day, and then move on to attack the Indian villages. The plan of this work he concerted in the evening with Major Ferguson, of the artillery. In the mean time, Colonel Oldham, an officer commanding the militia, was directed to send out that evening, two detachments, to explore the country and gain what knowledge they could of the enemy. The militia showed signs of insubordination, complained of being too much fatigued, and the order apparently could not be enforced. The militia were encamped beyond the stream, about a quarter of a mile in advance, on a high flat, a position much more favorable than was occupied by the main body. The placing of sentinels, about fifty paces from each other, formed their principal security against surprise.

At an early hour the next morning, the woods about the camp of the militia, swarmed with Indians, and a terrific yell, followed by sharp reports of the deadly rifle, were startling sounds, in the ear of the newly recruited soldier. The militia returned a feeble fire, and immediately fled toward the main body of the army. They came rushing in, pell-mell and threw into disorder the front rank, drawn up in the order of battle. The Indians, still keeping up their frightful yell, followed hard after the militia, and would have entered the camp with them, but the sight of troops drawn up with fixed bayonets to receive them, checked their ardor, and stopping short they threw themselves behind logs and bushes, and poured in a deadly fire upon the first line, which was soon extended to the second. Our soldiers were mown down at a fearful rate.

The Indians fought with great desperation. They charged upon the center of the two main divisions commanded by General Butler, and Colonel Darke with unexampled intrepidity. They aimed a destructive fire upon the artillerists from every direction, and swept them down by scores. The artillery if not very effective, was bravely served. A quantity of canister and some round shot were thrown in the direction whence the Indians fired; but concealed as they were, and seen only occasionally, as they sprang from one covert to another, it was impossible to direct the pieces to advantage; and so effective was the fire upon them, that every artillery officer, and more than two-thirds of the men, were killed or wounded.

St. Clair, unable to mount his horse, was borne about on a litter, and in the midst of peril and disaster, gave his orders with coolness and judgment. Seeing to what disadvantage his troops fought with a concealed enemy, he ordered Colonel Darke, with his regiment of regulars, to rouse the Indians from their covert with the bayonet, and turn their left flank. This was executed with great spirit; the enemy were driven three or four hundred yards; but for want of cavalry or riflemen, the pursuit slackened, and the troops were forced to give back in turn, and the Indians came on with a deadlier aim, the moment pursuit was relinquished. Strenuous efforts were made by the officers, early in the engagement, to restore order, which resulted in making themselves a mark, and they were cut down by the quick-sighted enemy.

All the officers of the Second regiment were cut off except three. The contest disastrous from the first, had now continued for more than two hours and a half. The loss of so many officers, and the hopeless condition of the army, the half of them killed, and the situation of the remainder desperate, brought discouragement to many a brave heart. It was useless to make further effort, which promised only a more fatal result. A retreat therefore was ordered, Colonel Darke being directed to charge the Indians that intercepted the way toward Fort Jefferson, and Major Clark with his battalion to cover the rear; these movements were successfully made, and the most of the troops that remained collected in a body, with such of the wounded as could possibly hobble along with them; thus they departed, leaving their artillery and baggage.

The retreat, though disorderly, was accomplished without difficulty, as the Indians did not pursue them far, from a desire to return for plunder. Yet the entire way, for near thirty miles, the distance to Fort Jefferson, bore the marks of a trepidation that seemed to characterize the entire engagement. The soldiers continued to throw away their guns, knapsacks, or whatever else impeded their flight, even when at a wide remove from all danger.

The army reduced by killed, wounded and desertion to about one-half its original number, fell back upon Fort Washington, the point of starting, and thus unfortunately closed a campaign, concerning which the highest expectations had been entertained. It was a heavy blow upon our infant republic, and spread over our country a gloom, which was greatly deepened by a sorrow for the loss of many worthy and brave men, who though they freely sacrificed their lives, could not avert these disasters.

The Indians, on account of this further victory, were elated beyond endurance, and conducted more haughtily than ever before. Their incursions were more frequent, their depredations more extensive, and their cruelties more excessive. The frontier inhabitants, especially of Pennsylvania, and Kentucky, never felt more insecure, and were never more exposed to loss of life, plunder and burning. In some instances whole settlements were broken up, by those who left their homes and sought, in the more densely peopled sections of the east, places of greater security.

These circumstances served to impart a deeper interest to the visit of a friendly deputation, consisting of about fifty chiefs of the Iroquois, who came to Philadelphia early in the spring of 1792, in compliance with the request of Colonel Pickering made at Painted Post the preceding year. Red Jacket was a prominent member of this delegation.

Their presence had been solicited, with the view of calling the attention of the leading chiefs, to thoughts and efforts for the improvement of their race; as well as by kind and generous treatment, to bring them into firmer alliance with the United States. And it is a pleasing thought that amid the wrongs done to the Indian, we are able to point to earnest and well intended endeavors, on the part of our government, to promote his welfare.

The governor of Pennsylvania cordially welcomed this deputation, representing the happiness their coming had created, and assuring them that every provision had been made, to render their stay agreeable, closing his remarks in these words:

"Brothers: I know the kindness with which you treat strangers that visit your country; and it is my sincere wish, that when you return to your families, you may be able to assure them, that the virtues of friendship and hospitality, are also practiced by the citizens of Pennsylvania."

To this welcome Red Jacket, a few days afterward replied, apologizing for not answering it sooner, and expressing the pleasure it afforded them, of meeting in a place where their forefathers in times past, had been wont to greet each other in peace and friendship, and declaring it as his wish, that the same happy relations might be established, and exist between the United States and all of the Indian tribes.

His remarks on peace were introduced by a beautiful reference to a picture of Penn's treaty with the Indians, and an enconium on the governors of Pennsylvania for their uniformly peaceable disposition.

It has been said of him as having occurred at a subsequent visit to the seat of our government, that when shown in the rotunda of the capitol, a panel representing, in sculpture, the first landing of the Pilgrims, with an Indian chief presenting them an ear of corn, in token of a friendly welcome, he exclaimed,—"*That was good.—The Indian knew they came from the Great Spirit, and he was willing to share the soil with his brothers.*"

When another panel was pointed out to him representing Penn's treaty,—he exclaimed sadly,—"*Ah! all's gone now.*" [Footnote: Drake's Book of the Indians.]

The Indians were again addressed by President Washington, who gave them a hearty welcome to the seat of government, declaring that they had been invited by his special request, to remove all causes of discontent, devise plans for their welfare, and cement a firm peace. He wished them to partake of all the comforts of the earth to be derived from civilized life, to be enriched by industry, virtue and knowledge, and transmit these invaluable blessings to their children.

The western Indians had charged the United States with an unjust possession of their lands. They desired no lands, he said, but such as had been fairly obtained by treaty, and he hoped the error might be corrected. For the further explanation of his views and wishes, he commended them to General Knox, the secretary of war, and Colonel Pickering; concluding his address with these words:—

"As an evidence of the sincerity of the desire of the United States for perfect peace, and friendship with you, I deliver you this white belt of wampum, which I request you will safely keep."

The president having thus appointed Colonel Pickering and General Knox, to attend to the further conferences with the Indians, Red Jacket's reply to the president's address, was made to them. His address was directed mainly to Colonel Pickering.

Taking in his hand the belt presented by President Washington, he spoke very much as follows:

"Your attention is now called to the words of the American Chief, when, the other day he welcomed us to the great council fire of the thirteen United States. He said it was from his very heart; and that it gave him pleasure to look around and see so large a representation of the Five Nations of Indians. That it was at his special request we had been invited to the seat of the general government, with a view to promote the happiness of our nation, in a friendly connection with the United States. He said also that his love of peace did not terminate with the Five Nations, but extended to all the nations at the setting sun, and it was his desire that universal peace might prevail in this land.

"What can we, your brothers of the Five Nations, say in reply to this part of his speech, other than to thank him, and say it has given a spring to every passion of our souls.

"The sentiment of your chief, who wishes our minds might all be disposed to peace,—a happy peace, so firm that nothing shall move it,—that it may be founded on a rock,—this comparison of the peace to a *rock, which is immovable,* has given joy to our hearts.

"The president observed also, that by our continuing in the path of peace, and listening to his counsel, we might share with you all the blessings of civilized life; this meets with our approbation, and he has the thanks of all your brothers of the Five Nations.

"And further, that if we attended to his counsel in this matter, our children and children's children, might partake of all the blessings which should rise out of this earth.

"The president observed again, that what he had spoken was in the sincerity of his heart, and that time and opportunity would give evidence that what he said was true. And we believe it because the words came from his own lips, and they are lodged deep in our minds.

"He said also that it had come to his ears that the cause of the hostilities of the western Indians, was their persuasion that the United States had unjustly taken their lands. But he assured us this was not the case; that none of his chiefs desired to take any of their land, without agreeing for it; and that the land, given up at the treaty at Muskingum, he concluded had been fairly obtained.

"He said to us that in his opinion the hostile Indians were in error, that whatever evil spirit, or lies had turned them aside, he wished could be discovered, that they might be removed. He had a strong wish that any obstacles to the extension of peace westward, might be discovered, so that they might be removed.

"In conclusion he observed that our professions of friendship and regard, were commonly witnessed by some token; therefore in the name of the United States, he presented us with this white belt, which was to be handed down from one generation to another, in confirmation of his words, and as a witness of the friendly disposition of the United States, towards the peace and happiness of the five confederated nations."

Red Jacket here laid down the white belt presented by the president, and taking up a belt of their own, continued his speech as follows:

"Now let the president of the United States possess his mind in peace. Our reply to his address to us the other day has been brief, for the belt he gave us is deposited with us, and we have taken firm hold of it. We return our united thanks for his address, in welcoming us to the seat of the great council, and for the advice he has given us.

"We have additional pleasure in knowing that you, Con-neh-sauty [Footnote: Col. Pickering.] are appointed to assist us, in devising the means to promote and secure the happiness of the Six Nations.

"Now open your ears, Representatives of the Great Council, Hear the words we speak. All present of the Great Council, [Footnote: Referring to members of Congress present.] and our brethren of the Five Nations, hear! We consider ourselves in the presence of the Great Spirit, the proprietor of us all.

"The president has in effect told us we are freemen, the sole proprietors of the soil on which we live. This has gladdened our hearts, and removed a weight that was upon them. This indeed is to us an occasion of joy, for how can two brothers speak freely together, unless they feel they are upon equal ground?

"We now speak freely, as they are free from pressure, and we join with the president in his wish, that all the evils which have hitherto disturbed our peace, may be buried in oblivion. This is the sincere wish of our hearts.

"Now, Brother, continue to hear, let all present open their ears, while those of the Five Nations here present speak *with one voice*. We wish to see your words verified to our children, and children's children. You enjoy all the blessings of this life; to you therefore we look to make provision, that the same may be enjoyed by our children. This wish comes from our heart, but we add that our happiness cannot be great if, in the introduction of your ways, we are put under too much constraint.

"Continue to hear. We, your brothers of the Five Nations, believe that the Great Spirit let this island [Footnote: The Indians use the term *island,* in speaking of this continent.] drop down from above. We also believe in his superintendency over this whole island. He gives peace and prosperity, he also sends evil. Prosperity has been yours. American brethren, all the good which can spring out of this island, you enjoy. We wish, therefore, that we and our children, and our children's children, may partake with you of that enjoyment.

"I observe that the Great Spirit might smile on one people, and frown on another. This you have seen, who are of one color, and one blood. The king of England, and you Americans strove to advance your happiness by extending your possessions on this island, which produces so many good things. And while you two great powers were contending for those good things, by which the whole island was shaken, violently agitated, is it strange that our peace, the peace of the Five Nations, was shaken and overthrown?

"But I will say no more of the trembling of this island. All in a measure is now quiet. Peace is restored. Our peace, the peace of the Five Nations is beginning to bud forth. But still there is some shaking among our brethren at the Setting Sun; and you, of the thirteen fires, and the king of England know what is our situation and the cause of this disturbance. Here now, you have an ambassador, [Footnote: Referring to the British envoy to the United States.] as we are informed from the king of England. Let him in behalf of the king, and the Americans, adjust all their matters, according to their agreement, at the making of peace—and then you will soon see all things settled among the Indian nations. Peace will extend far and near. Let the president and the ambassador use all their exertions to bring about this settlement, according to the peace, and it will make us all glad, and we shall consider both as our real friends.

"Brother: Continue to hear! Be assured we have spoken not from our lips only, but from our very hearts. Allow us then to say: That when you Americans and the king made peace, he did not mention us, showed us no compassion, notwithstanding all he said to us, and all we had suffered. This has been the occasion to us, the Five Nations, of great loss, sorrow and pain. When you and he settled the peace between you two great nations, he never asked for a delegation from us, to attend to our interests. Had this been done, a settlement of peace among all the western nations might have been effected. But neglecting this, and passing us by unnoticed, has brought upon us great pain and trouble.

"It is evident that we of the Five Nations have suffered much in consequence of the strife between you and the king of England, who are of one color and of one blood. But our chain of peace has been broken. Peace and friendship have been driven from us. Yet you Americans were determined not to treat us in the same manner as we have been treated by the king of England. You therefore desired us at the re-establishment of peace, to sit down at our ancient fireplaces, and again enjoy our lands. And had the peace between you and the king of England been completely accomplished, it would long before this have extended far beyond the Five Nations.

"BROTHER CON-NEH-SAUTY: We have rejoiced in your appointment, for you are specially appointed with General Knox, to confer with us on our peace and happiness. We hope the great warrior will remember, that though a *warrior*, he is to converse with us about *peace*; letting what concerns war sleep; and the counselling part of his mind, while acting with us, be of *peace*.

"Have patience, and continue to listen. The president has assured us that he is not the cause of the hostilities now existing at the westward, but laments it. Brother, we wish you to point out to us of the Five Nations, *what you think is the real cause.*

"We now publicly return our thanks to the president, and all the counsellors of the thirteen United States, for the words he has spoken to us. They were good, unqualifiedly good. Shall we observe that he wished that if the errors of the hostile Indians could be discovered, he would use his utmost exertions to remove them?

"BROTHER! You and the king of England are the two governing powers of this island. What are we? You both arc important and proud; and you cannot adjust your own affairs agreeably to your declarations of peace. Therefore the western Indians are bewildered. One says one thing to them, and another says another. Were these things adjusted, it would be easy to diffuse peace everywhere.

"In confirmation of our words, we give this belt, which we wish the president to hold fast, in remembrance of what we have now spoken." [Footnote: This speech, given by Col. Stone from a manuscript of J. W. Moulton, Esq., on account of its importance, is presented almost entire. A few changes have been made, but the ideas of the orator, and the language mostly in which they are given, have been strictly maintained, while the changes are no greater than would have been made, had two reporters taken the words as they came from the lips of the orator.]

A very touching reference is made in this speech, to the manner in which the Indians had been treated by Great Britain, when peace was concluded with the United States. Notwithstanding the promises and high expectations held out to them, at the commencement of the war, and their sacrifices and services during its continuance, no notice was taken, no mention made of them in the treaty of peace. In the expressive language of Red Jacket, *"the king showed them no compassion."* They had for years fought side by side with the soldiers of Britain, they had, with stealthy tread, come down upon our settlements far removed from the seat of war, surprised peaceful inhabitants, slain defenseless women and children, plundered and burned their dwellings, and wrought in the hearts of the American people a sense of wrong, that cried for redress. What could be their position, now that the armies of Britain are withdrawn? The armies of Britain defeated, could they, single handed, cope with the American army? These were questions that weighed deeply on their minds. Did they expect the hand of friendship to be extended toward them? To be invited to councils of peace, —to the intimacies, hospitalities, and kindly feeling manifested on this occasion? The orator was deeply impressed by it, and notes the contrast apparent in the conduct toward them, of Britain and America. *"You Americans were determined not to treat us in the same manner, as we had been treated by the king of England. You desired us at the re- establishment of peace, to sit down at our ancient fire-places, and again enjoy our lands."* He further very significantly refers to the occasion of the hostile feelings among the Indians at the West. It was because the peace between England and America *"had not been fully accomplished."* In other words, hostile feelings were still cherished, and their *outward manifestation* could be seen, in the plundering and massacres, still carried on among our frontier settlements. The establishment of a *true peace* between the two countries,—the existence and cultivation of genuine amicable relations between them, would, in his view, end all this trouble, and *"diffuse peace everywhere."*

We have already had occasion to notice the unfriendly feeling, cherished by the British Indian Department in Canada, toward the United States; and evidence will be afforded further on, of their being deeply implicated in the hostilities endured, coming from the Indians on our western border.

CHAPTER IX

Indian appropriation—Embassy sent West—Instructions—Medal presented to Red Jacket—Military suits—Close of conference—Washington's parting words—Visit of Thayendanegea—Council at Au Glaize—Result—Another commission—Indian diplomacy—Washington's letter to Mr. Jay—Commission goes West—Various interviews—Result of council—Re-organization under General Wayne—Ready for action—Advice of Little Turtle—Wayne's battle and victory—Treaty of peace.

While these Indian chiefs were at Philadelphia, a bill was passed by Congress, and ratified by the president, appropriating fifteen hundred dollars annually, for the benefit of the Iroquois, in purchasing for them clothing, domestic animals, implements of husbandry, and for encouraging useful artificers to reside in their villages.

They were engaged also to go on a pacific embassy to the hostile Indians of the West, and assure them of the friendly disposition of the United States toward them;—that they want nothing which belongs to the hostile Indians;—that they appointed commissioners to treat with them for their lands, and give them a large quantity of goods;—that a number of chiefs signed the deeds, and from the reports of the commissioners, it was supposed the lands had been fairly obtained;—that under this supposition large tracts had been sold, and hence difficult to restore again; but as the United States desire only what is just, they will attentively hear the complaints of the western Indians;—they will re-examine the treaties, and inquire into the manner in which they were conducted;—and if the complaints of the western Indians, appear to be well founded, the United States will make them ample compensation for their lands. They will do more;—so far from desiring to injure, they would do them good; they would cheerfully impart to them that knowledge, and those arts, by which they propose to increase the happiness, and promote the welfare of the Six Nations.

It was during this visit that President Washington, in token of his friendship and esteem, gave Red Jacket a large silver medal bearing his likeness, which he ever after preserved with much care, and took great pride in wearing.

General Knox, the secretary of war, directed also that a military suit of clothes be given to each member of the deputation, including a cocked hat, as worn by the officers of the United States army. When Red Jacket's suit was presented to him he eyed it carefully, and rather admiringly, but requested the bearer to inform General Knox that the suit would hardly become him, as he was not a war-chief but a sachem, the sachems being civil, rather than military officers. He desired therefore that another suit be prepared, which would accord better with the relation he sustained to his people; at the same time declaring the one sent very good, and manifesting a disposition to retain it, until the other was prepared. A plain suit was accordingly prepared and brought to him, and with this he seemed to be highly pleased. The bearer tarrying a little, and manifesting a readiness to carry back the other suit, Red Jacket coolly and rather playfully remarked, that though the present suit was more in keeping with his character as sachem, it nevertheless, occurred sometimes, in cases of emergency, that the sachems also went to war, and as it would then be very becoming and proper for him to wear it, he was happy to have one in case a circumstance of this kind should occur.

These Indian chiefs were all highly gratified with the attention shown them, during this visit to the general government. They were especially pleased with the interest that had been taken in the improvement of their people, and the pledges they had received of aid in carrying out the benevolent designs entertained toward them. And they all, Red Jacket with the rest, were favorably impressed with the views of Washington, in desiring to introduce among them the improvements of civilized life.

These conferences were brought to a close on the thirtieth of April, and President Washington in a concluding speech, said to them,—"When you return to your country, tell your nation that it is my desire to promote their prosperity, by teaching them the use of domestic animals, and the manner that the white people plough and raise so much corn, and if upon consideration, it would be agreeable to the nation at large, to learn those arts, I will find some means of teaching them at some places within their country, as shall be agreed upon." [Footnote: Irving's Life of Washington.]

The government had taken special pains also to secure the attendance of the celebrated Thayendanegea or Brant, with this deputation of friendly chiefs. The invitation, though a pressing one, was declined, and not without reason. For besides the powerful influence exerted over him by the officers of the British government in Canada, who strenuously opposed his coming, it has since been ascertained that he was the leading spirit who directed with so much success to the Indians, the onslaught upon General St. Clair's army, the preceding fall. Hence his own feelings could not have been of the most friendly character. He was, nevertheless, induced to visit the seat of government during the month of June following, and pledged himself to exert his influence in an effort to secure peace for the United States, with the Indians at the West.

A very large Indian council, composed of delegates from many and some of them very distant nations, was held at Au Glaize, on the Miami of Lake Erie, in the autumn of 1792. A large delegation from the Six Nations, friendly to the United States, was present and took part in the deliberations. Red Jacket was the principal speaker, and strenuously advocated the settlement of their difficulties, by peaceful negotiations instead of war.

The Shawanese as strenuously advocated the continuance of hostilities. They taunted the Six Nations with having induced them to form a great confederacy, a few years before, and of having come to the council now, "with the voice of the United States folded under their arm;"—referring to the belt which was significant of their embassy.

The Shawanese, Miamis and Kickapoos were addicted to horse-stealing, and while hostilities were continued, they reaped from this source, their greatest harvests.

Captain Brant on account of sickness was unable to be present, and it may be noticed that from this time on, his efforts to form a North-western Indian Confederacy, were very sensibly remitted. He no doubt found there were so many conflicting interests and national jealousies in the way, as to render the project comparatively hopeless. But more than all, he had depended upon the following of the entire body, composed of the Six Nations, and when he saw them coming largely under the influence of the United States, he could realize that the strength and permanence of his contemplated position, were so seriously affected, as to render its attainment extremely doubtful. The addition of the entire Iroquois family, to the proposed confederation, would have brought into it an element of intellectual superiority, and their long established polity of acting in concert, would have been of essential service among forces that were wild and chaotic. And we are not surprised that the diversion effected among them, should have changed somewhat the views of the distinguished Thayendanegea.

No decisive action was reached at this council, but an agreement was made to suspend hostilities during the winter, provided the United States would withdraw their troops from the west side of the Ohio; and another council was appointed to meet at the Miami Rapids during the following spring.

The Iroquois delegation forwarded to our government a report of the service they had rendered, the action taken by the council, and the agreement to meet in the spring, and requested that agents might be sent, "who were men of honesty, not land-jobbers, but men who love and desire peace. We also desire that they may be accompanied by some Friend, or Quaker, to attend the council."

On the 19th of February, 1793, General Benjamin Lincoln, Beverly Randolph and Colonel Pickering were commissioned by the president to attend the great Indian council at Miami Rapids, in the ensuing spring.

Meanwhile the Indians, dissatisfied with the views of the president, as transmitted by the Six Nations, held another council at Au Glaize in February, and framed a very explicit address to the Six Nations, affirming they would listen to no proposition from the United States, that did not concede the Ohio river, as the boundary line between them, and the Indian country. They desired the United States to be fully apprised of this, before sending their delegation; and they notified the Six Nations of a private council at Miami Rapids, before meeting the American commissioners, to adjust their opinions, so as to speak but one language at the council; they further declared their intention not to meet the commissioners at all, until assured they had authority to conclude a treaty on this basis.

In this determination they were encouraged, and sustained by the British Indian Department of Canada. President Washington, in a letter to Mr. Jay, our minister in London, writing in 1794, very clearly sets forth the work thus accomplished.—He says:—"There does not remain a doubt, in the mind of any well informed person in this country, not shut against conviction, that all the difficulties we encounter with the Indians, their hostilities, the murder of helpless women and children, along our frontiers, result from the conduct of agents of Great Britain in this country. In vain is it then for its administration in Britain to disavow having given orders which will warrant such conduct, whilst their agents go unpunished; while we have a thousand corroborating circumstances, and indeed as many evidences, some of which cannot be brought forward, to prove that they are seducing from our alliances, and endeavoring to remove over the line, tribes that have hitherto been kept in peace and friendship at great expense, and who have no causes of complaint, except pretended ones of their creating; whilst they keep in a state of irritation the tribes that are hostile to us, and are instigating those who know little of us or we of them, to unite in the war against us; and whilst it is an undeniable fact, that they are furnishing the *whole with arms, ammunition, clothing, and even provisions to carry on the war,* I might go farther, and if they are not much belied, add, *men in disguise.*" [Footnote: Marshall's Washington.]

The commissioners of the United States appointed to confer with the Indian tribes at the West, proceeded on their way, arriving at Niagara the latter part of May, 1793. Here they were very kindly entertained by Governor Simcoe until the council was ready to receive them.

While here they were visited by a large deputation from the council at Miami Rapids, who desired an explicit answer to the inquiry whether they were authorized to run and establish a new boundary? Which they answered in the affirmative, at the same time reminding the Indians that in almost all disputes there were wrongs on both sides, and that, at the approaching council, both parties must expect to make some concessions.

This reply was well received and sanguine hopes were entertained of a favorable termination of their mission.

The Indians returned again to their council at Miami, and the commissioners supposing they would now be prepared to receive them, proceeded on their voyage westward. Arriving at the mouth of Detroit river they were obliged to land, being forbidden by the British authorities to proceed any farther toward the place of meeting.

They were met here by another Indian deputation, bringing a paper with a written statement of their determination, to make the Ohio the boundary line between the Indian country and the United States, and requiring the latter, if sincere in their desires for peace, to remove their settlements to the south side of that river. To this the commissioners were desired to give an explicit written answer.

They replied, referring to the understanding from their conference at Niagara, that some concessions were to be made on both sides, and giving a brief history of the treaties by which a title had been acquired to land north of the Ohio, on the faith of which, settlements had been formed which could not be removed; hence they answered explicitly.—"*The Ohio river cannot be designated as the boundary line.*"

They expressed the hope that negotiations might proceed on the basis of these treaties, closing with some concessions, and liberal offers for some lands still held by the Indians.

The debate at this council, it is said, ran high. Thayendanegea, and others of the Six Nations were strenuous in their advocacy of peace. The offer of the commissioners to establish a boundary line that would include the settlements already made north of the Ohio, they regarded as reasonable, and that farther concessions ought not to be required. Quite a number of tribes were influenced to adopt this view, which at one time it was thought would prevail. But there were certain ruling spirits present determined to make no concession, and the council broke up without allowing the commissioners, or any other white person, not in sympathy with Britain, to be present.

Previous to the holding of this council, the army had been re-organized under the command of General Anthony Wayne, an officer of untiring energy and vigilance; a larger number of soldiers had been called into the field, and as they were placed under a severe discipline, to inure them to the dangers and hardships of the campaign, it was undertaken with flattering prospects of success.

Pittsburgh had been made the place of rendezvous; but fearing the influence of an encampment near a town, and wishing to inspire in his soldiers a feeling of self reliance, General Wayne, on the 27th of November, 1792, marched his army to a point twenty-two miles distant on the Ohio, which he called Legionville, fortifying it and taking up his quarters there for the winter.

On the 30th of April, 1793, as spring had opened, he broke up his garrison at Legionville, and led his army down the river, to Fort Washington, its site being that of the present beautiful and flourishing city of Cincinnati.

Here he remained while the negotiations were going on with the Indians at the West. As soon as they were ended and the result known, he took a more advanced position, marching in October in the direction pursued by, General St. Clair, to a point on the south-west branch of the Miami, six miles beyond Fort Jefferson, and eighty from Fort Washington, which he fortified and called Greenville.

On the 23d of December, a detachment of the army commanded by Major Burbeck took possession of the ground where the army of General St. Clair, two years before on the 4th of the preceding November, had sustained a terrible defeat. Here they gathered up sadly and sacredly the bones that marked this as a place of human slaughter, put in order the field-pieces that were still upon the ground, served them with a round of three times three, over the remains of their fallen comrades, and erected a fortress, appropriately naming it Fort Recovery.

The army at different points had skirmishes with the enemy that were not serious, but they served to create confidence and inspire courage in the minds of the soldiers.

It was not until the 20th of August, 1794, that General Wayne had a regular engagement with the Indians. Yet like a true gladiator he had been preparing for the struggle, and his wariness, which had gained for him the title of "*Black Snake*" may be gathered from the speech of Little Turtle, chief of the Miamis, and one of the most active and brave warriors of his time.

He counselled his countrymen to think favorably of the proposals of peace offered by General Wayne before giving them battle; saying,—"We have beaten the enemy twice under separate commanders. We cannot expect the same good fortune always to attend us. The Americans are now led by a chief who never sleeps. The night and the day are alike to him; and during all the time he has been marching on our villages, notwithstanding the watchfulness of our young men, we have never been able to surprise him. There is something that whispers to me,—*it would be prudent to listen to his offers of peace.*"

But this counsel was rejected by the Indians, who determined to give battle to the Americans the next day. They fought in the vicinity of a British fort, which Governor Simcoe of Canada had caused to be erected at the foot of the rapids of the Miami emptying into the lakes, far within the acknowledged territory of the United States.

The ground occupied by the Indians was well chosen, being a thick wood, where were old fallen trees that marked the track of some ancient hurricane, where the use of cavalry would be impracticable, a place suited to afford them shelter and well adapted to their peculiar mode of warfare. But the order of General Wayne to advance with trailed arms, and rouse the Indians from their covert at the point of the bayonet, and when up deliver a close and well directed fire on their backs, followed by a brisk charge, so as not to give them time to load again; was executed so promptly, and with so much effect that the Indians were driven in one hour more than two miles, and soon dispersed in terror and dismay, leaving the ground in full and quiet possession of the victorious army.

This battle, which terminated within reach of the British guns, decided the fate of the campaign. The Indians after this were dispirited and unable to make a general rally. The distrust awakened by the coolness of their supposed friends, the gates of whose fort remained unopened while they were fleeing thither for a covert, served not less than the victory to dishearten them, and incline their thoughts toward peace.

The few days spent by the army on the battle ground after its victory, were occupied in destroying the property of the Indians in that vicinity, including also the extensive possessions of Colonel McKee, an officer of the British Indian Department, whose influence had been exerted in promoting these hostilities, whose effects were now being experienced. The fort itself was poised in the General's mind, as was also the torch of the gunner, who was only restrained by his commanding officer from firing upon Wayne, who, as he thought came too near, in making his observations on one of His

Majesty's forts. Prudence prevailed. The fighting was confined to a war of words in a spirited correspondence between General Wayne, and the officer in command of the fort.

General Wayne after laying waste their principal towns in this region, continued in the Indian country during the following year, bringing his campaign to a close by a treaty with the North-western tribes, which was entirely agreeable to the wishes of the United States.

CHAPTER X

Canandaigua at an early day—Facts in the early settlement of Bloomfield—Indian Council—Its object—Indian parade—Indian dress—Opening of Council—Speeches—Liberal offers of the government—Mr. Savary's Journal—Treaty concluded—Account of Red Jacket by Thomas Morris.

Canandaigua at an early day was the objective point for all who were seeking what was called the Genesee country. It was at the head of navigation. Parties coming from the east could transport their goods by water from Long Island Sound to Canandaigua, with the exception of one or two carrying places, where they were taken by land.

We can hardly realize that at that time there was here a widely extended forest, in all its loneliness and grandeur. Its first trees were cut down in the fall of 1788, soon after Mr. Phelps had concluded his treaty of purchase with the Indians. By means of them a log store-house was constructed, near the outlet of the lake. The family of a Mr. Joseph Smith took possession of it in the spring of 1789. Judge J. H. Jones, who in the fall of 1788, was one of a party to open a road between Geneva and Canandaigua, witnessed, on revisiting the latter place in 1789, a great change.

"When we left," he says, "in the fall of '88, there was not a solitary person there;—when I returned fourteen months afterwards, the place was full of people; residents, surveyors, explorers, adventurers; houses were going up; it was a thriving, busy place." During the following year quite a nucleus for a town had gathered here. In 1794, Mrs. Sanborne, an enterprising landlady, whose eye kindled with the recollection of those days, served up in a tea saucer the first currants produced in the Genesee country. [Footnote: Conversation of the author with Mrs. Sanborne.] Canandaigua at that time and for many years after was head-quarters for all who were making their way into what at that time was called the Indian country, and from the respectability and enterprise of its early inhabitants, it became attractive as a place of residence.

But though considerable improvements had been made here, the entire region was new, romantic and wild. Such was its condition at the time of the great Indian council that convened here in the autumn of 1794. Indians and deer, and wolves, and bear were very abundant and were mingled with the early associations of those who contributed to make this an abode of elevation and refinement. The cow-boy, often startled while on his way by the appearance of a bear, went timidly forth on his evening errand, inspired with courage by the thought that he might, for his protection, shoulder a gun. Bear incidents, narrow escapes from fighting with bears, and bear stories of every description, entered largely into the staple of their conversation, and many an evening's hour was thus beguiled away, around the huge and brightly blazing fire of the early pioneer.

"Did you hear," said a Mrs. Chapman to a Mrs. Parks, how neighbor Codding came near being killed yesterday?

"Mercy! no. How did it happen?

"Mr. Codding was in the woods splitting rails, and just as he was turning around to take up his axe to cut a sliver, don't you believe he saw a great bear sitting up on his hind legs, and holding out both fore paws ready to grab him."

"Mercy on us! What did he do?

"What did he do? He took up his axe, and instead of cutting the sliver, cut into the old bear's head. But the axe glanced and only cut into the flesh, without killing the bear, and he ran away with the axe sticking fast in the wound.

"Awful! Awful! How thick the bears are getting to be! Husband says they have killed off most all of our hogs.

"Your hogs! Just think once, there was a great bear came the other night and got hold of a hog in Asahel Sprague's hog-pen, and would have killed him, if Mr. Sprague hadn't shot the old fellow.

"Yes, and last summer when Mr. Sperry was gone off to training, there was a bear came in the day time and tackled one of their hogs right in their own door yard; but Mrs. Sperry and the children screamed so awfully, and gave him such a tremendous clubbing, he was glad to put off into the woods again.

"Ha! Ha! She was about up to Jim Parker, who broke a bear's back with a hand-spike in driving him out of his corn field, just as he was climbing over the fence." [Footnote: Facts which transpired in the early history of Bloomfield. See Turner's History.]

Wolves were equally if not more numerous, destroying in some instances entire flocks of sheep, so that there was not a farmer in the region who did not suffer more or less from their depredations.

It was something of an off-set to these annoyances that deer were very abundant, and furnished the inhabitants with an ample supply of their delicious meat. The Indians while assembled here during the council, often killed more than a hundred of them in a single day.

The object of convening this council was to settle difficulties of long standing, and quiet the minds of the Iroquois, who were much disturbed by the warlike spirit prevailing at this time among the Indians at the West. The influences from this source were of such a nature as to render many among these friendly tribes exceedingly bold. In some instances on entering the houses of settlers they would manifest a very haughty temper, and rudely demand a supply of their wants as though they were still proprietors and lords of the soil, and the settlers only their servants or tenants.

The settlers themselves began to feel unpleasantly about their position. During the spring of this same year while Thomas Morris was painting his house, erected the previous summer, and making other improvements around it, indicating his design of having a permanent and inviting home, it so happened that a company of settlers in passing by, paused to view with astonishment what was going on. From a feeling of insecurity they had just abandoned their new locations in this region, and had come thus far on their way, having resolved to return to the more safe and quiet homes they had left at the east. But beholding the enterprise of Mr. Morris, and the business and thrift that prevailed here on every side, they inferred that their situation could not be so very precarious, and wisely concluded to return and carry forward the improvements commenced by themselves.

The Indian council, held during the months of October and November, had been appointed before the victory of General Wayne, noticed in the preceding chapter, had transpired. This had much to do in giving a favorable turn to the proceedings, and of securing those pacific relations with the Iroquois, that were then established. Before this these tribes and the Indians generally were stimulated with the idea that they might form and maintain in the North-west an independent nationality, that would reflect once more the pride and glory of the ancient dominion of the Iroquois. But when the news of this signal victory was circulated among them, their spirits were humbled and broken. They seemed to relinquish this dream of greatness, and gave themselves up to the stern demands of an evident necessity. This sad intelligence, however, did not reach them until the council had been for several days in progress. Its first opening was darkened by no cloud of evil. There was nothing to hinder the exercise of that proud bearing with which their past greatness, and a hopeful future inspired them.

They began to assemble by the arrival of the Oneidas on the eleventh of October. The Onondagas, Cayugas, and a part of the Senecas, led by Farmer's Brother, came in on the fourteenth. Cornplanter at the head of the Allegany clan of Senecas arrived on the sixteenth, and Red Jacket with his, on the eighteenth.

On assembling, a degree of dignity and decorum was manifested, which served to indicate their ideas of the forms and proprieties due to the occasion. Before reaching the council fire the chiefs and warriors halted, carefully decorated themselves after their manner, and then marched to meet those appointed to confer with them on the part of the government, and after passing around and encircling them, with the train, the leader stepped forth, formally announced their arrival in obedience to the summons they had received, at the same time delivering the belt brought by the messenger sent to call them together.

The next tribe that came, halted and prepared themselves as the others, were received by the tribe or tribes already on the ground, who also arrayed themselves in their uniform, and having received their welcome, salutes being fired and returned, they marched all together and formed in a circle around the commissioners, when the same ceremony was observed, as before, of delivering the belt. They proceeded thus until all the Indians had assembled to the number of about sixteen hundred.

It was an occasion for the display of Indian pageantry, and though it may have been more rude than among nations calling themselves civilized, it was the same in its essential elements, and this council was ushered in with as true a military spirit as though banners had been flying, bayonets gleaming, and soldiers marching to the liveliest, or most heart-stirring sounds of music.

The uniform of the Indian was not as the dress of the European, ornamented, epauletted, tinselled; it was a more simple, less expensive, but not a less time honored mode of adorning his person. Though his military coat was of paint of different colors with which he was striped in a distinguishing manner, he regarded it no doubt as gorgeous and gay. Instead of the gracefully waving plume he was bedecked with the feathers of the kingly eagle; beads and shells served in the place of military buttons; and his trophies in the chase, and in war, he regarded as forming a prouder sash than the richest scarf of scarlet or of blue.

Canandaigua, in years gone by, has often witnessed scenes of proud military display. But never will there be witnessed so grotesque, and in many respects so imposing a parade as appeared on this occasion. The neighboring forest swarmed with life, and resounded with the wild yell and deafening war-whoop of the Indian. It was his gala day, and highly fitting that before surrendering these grounds forever to the dominion and usages of another race, he should come forth once more from his native wilds, and depart in the fullness of his strength, as the sun passing from under a cloud, sheds his full glory over the earth before sinking beneath the western horizon. This was his last day of pride on ground hallowed in the memories of the past.

The occasion called forth an unusual attendance. It was known that Colonel Pickering who had been appointed to hold this treaty, would come prepared to give them a grand feast, and distribute among them a large amount of money and of clothing. Hence they all came. "For weeks before the treaty, they were arriving in squads from all their villages, and constructing their camps in the woods, upon the lake shore, and around the court-house square. The little village of whites was invested, overrun with the wild natives. It seemed as if they had deserted all their villages, and transferred even their old men, women and children to the feast, the carousal, and the place of gifts. The night scenes were wild and picturesque; their camp fires lighting up the forest, and their whoops and yells creating a sensation of novelty not unmingled with fear, with the far inferior in numbers who composed the citizens of the pioneer village and the sojourners of their own race." [Footnote: History of the Phelps and Gorham Purchase.]

The council was formally opened on the eighteenth of October, by a speech of condolence on the part of the Oneidas and Onondagas, to the Senecas, Cayugas, Tuscaroras and Delawares, some of whom were present, on account of the death of a number of their chiefs since the last meeting.—It was with a view to *"wipe away their tears,—brighten their faces, and clear their throats,"*—that they might speak freely at the council fire.

Red Jacket in reply made a very sympathetic, and as it was regarded at the time, beautiful address, presenting belts and strings of wampum to *"unite each to the other as the heart of one man."*

Next was given a speech of congratulation by Colonel Pickering, who appointed a council of condolence on the following Monday for the Delawares, who were mourning for a young brother killed by a white man.

The ceremony of burying the dead,—covering the grave with leaves to obscure it from sight,—of burying the hatchet taken from the head of the victim, thus representing his death by violence,—of covering it with stones and pulling up and planting over it a pine tree, so that in after years it should never be disturbed; of wiping the blood from the head of the victim, and tears from the eyes of the mourners,—these things represented by speech and action having been performed, the council was opened in earnest on the day following.

In reply to Colonel Pickering's remarks of the preceding day respecting peace, and upon keeping the chain of friendship bright, Fish Carrier, an aged and influential chief, in a speech of some length recounted the history of the whites and of their intercourse with the red men from their first settlement in this country. He referred to the manner in which they had been received, to the friendship, that had existed before the controversy of the United States with Great Britain, and to the negotiations that had taken place since that time, the grievances they had suffered, dwelling particularly upon the dissatisfaction still existing among them about the treaty at Fort Stanwix in 1784. "The commissioners were too grasping, they demanded of us too much." But as they had taken hold of the chain of friendship with the fifteen fires they were disposed to hold fast; but he thought it needed brightening up a little.

Colonel Pickering in reply to them said they ought not to think very hard of them about that treaty, for they had just come out of a long and bloody war, and as they had been victorious the Six Nations ought not to blame them for feeling a little proud; and they ought not to be surprised, in view of what had taken place during the war, if the commissioners were somewhat severe.

A deputation from the Quakers was present at this council, and their address being read and interpreted, was received by the Indians with much satisfaction.

At the opening of the council the next day, a request was presented by some of the Indian women, who desired that their views might be heard; and permission being granted, Red Jacket spake in their behalf.

He represented that the women had taken a deep interest in everything pertaining to the wellfare of their nation; and he requested the sachems and warriors, as well as the commissioner, to give an attentive consideration to the views of those he had been chosen to represent. They had attended upon the council, had listened to all that had been said, and they desired it to be understood that their views were in accordance with those of their sachems and chiefs. They felt that the white people had caused them a great deal of suffering. The white people had pressed and squeezed them together until their hearts were greatly pained, and they thought the white people ought to give back all their lands. A white woman had told the Indians to repent; [Footnote: Referring to Jemima Wilkinson.] they wished in turn to call upon the white people to repent; they needed to repent as much as the Indians;

and they hoped the white people would repent and not wrong the Indians any more.

The commissioner thanked them for their speech, saying he had a high respect for the women, and would be happy to hear from them whenever they had anything to say.

After several days had passed without coming to anything decisive in regard to the main object of the council; Colonel Pickering called their attention to the fact of their grievances, saying they had been together sometime and talked them over and had found but two rusty places in the chain of friendship, one of which they had already brightened. But the other spot they thought was too deep to be cleared up. It related to their lands. He then showed them maps which clearly pointed out the limits agreed upon in their treaties, and by a distinct statement of the negotiations and treaties that had been made at different times with them, and afterward confirmed, proving that the claims of the United States were just, he declared himself ready to stipulate concerning their grievances, that they should still have the privilege of hunting upon the lands they had ceded, and that their settlements thereon should remain undisturbed. He further assured them that the United States would increase their annuity from fifteen hundred to four thousand five hundred dollars, to say nothing about the presents he had brought them amounting to ten thousand dollars. These he would distribute in case of a favorable termination of their council. He hoped in view of these liberal offers they would dismiss their complaints, bury the hatchet deep and take hold of the chain of friendship so firmly as never again to have it torn from their grasp.

The Indians appeared to be pleased with these offers, and promising to regard them favorably, spent several days in deliberating among themselves, inviting to their councils the Quakers, a deputation of whom, as we remarked were present. William Savary, one of their number made the following interesting note of his observations at the time.—

"Oct. 30. After dinner John Parrish and myself rode to view the Farmer's Brother's encampment which contained about five hundred Indians. They are located by the side of a brook in the woods: having built about seventy or eighty huts, by far the most commodious and ingeniously made of any I have seen. The principal materials are bark, and boughs of trees, so nicely put together as to keep the family dry and warm. The women as well as the men appeared to be mostly employed. In this camp there are a large number of pretty children, who in all the activity and buoyancy of health, were diverting themselves according to their fancy. The vast number of deer they have killed, since coming here, which they cut up and hang round their huts inside and out to dry, together with the rations of beef, which they draw daily, give the appearance of plenty to supply the few wants to which they are subjected. The ease and cheerfulness of every countenance, and the delightfulness of the afternoon, which the inhabitants of the woods seemed to enjoy with a relish far superior to those who are pent up in crowded and populous cities, all combined to make this the most pleasant visit I have yet made to the Indians; and induced me to believe that before they became acquainted with the white people, and were infected with their vices, they must have been as happy a people as any in the world. In returning to our quarters we passed by the Indian council, where Red Jacket was displaying his oratory to his brother chiefs on the subject of Colonel Pickering's proposals."

Mr. Savary again observes:—"Red Jacket visited us with his wife and five children, whom he had brought to see us. They were exceedingly well clad, in their manner, and the best behaved and prettiest Indian children I have ever met with." [Footnote: As quoted by Col. Stone.]

Various councils and deliberations with the Indians, resulted finally in the conclusion of a treaty, which was quite satisfactory to all the parties.—By this treaty peace was again declared to be firmly established, the different tribes were confirmed in their reservations, and lands that had not been sold, the boundaries of which were accurately described, and the United States engaged never to claim these lands, or disturb the Six Nations in the free use and enjoyment of them. The Six Nations pledged themselves also not to claim any other lands within the boundaries of the United States, nor disturb the people of the United States in the free use and enjoyment thereof. It was stipulated also that the United States should have the right of way for a public road from Fort Schlosser to Lake Erie, have a free passage through their lands, and the free use of harbors and rivers adjoining and within their respective tracts of land, for the passing and securing of vessels and boats, and liberty to land their cargoes, where necessary for their safety.

In consideration of these engagements the United States were to deliver the presents, and pay the annuity as already intimated in the promise of Colonel Pickering.—The money thus pledged was to be expended yearly forever in purchasing clothing, domestic animals, implements of husbandry, and other utensils suited to their circumstances, and for compensating useful artificers who might be employed for their benefit.

It was further agreed that for injuries done by individuals on either side, private revenge should not take place, but that complaint be made by the injured party to the nation to which the offender belonged, and that such measures were then to be pursued as should be necessary for the preservation of peace and friendship. [Footnote: Indian Treaties. Favored with a copy by O. Parrish. Esq., of Canandaigua, N. Y.]

The conclusion of this treaty was regarded as a great point gained. Previous to this time, such of the Iroquois as remained in their ancient seats, were but partially reconciled to the United States, and were oscilating in their friendship. But henceforth they were uniformly steadfast in the allegiance they had promised.

The holding of this council was further useful in withdrawing the attention of this large body of Indians with their warriors, who had been earnestly solicited to join their hostile brethren at the West.

During the progress of the council there were several speeches made, but as they are not of special interest or importance they have not been given. Colonel Stone mentions an evening when quite a number of the chiefs dined with Colonel Pickering. He says,—"Much good humor prevailed on this occasion. The Indians laid aside their stoicism, indulged in many repartees, and manifested the keenest relish for wit and humor. Red Jacket, in particular, was conspicuous for the readiness and brilliance of his sallies." [Footnote: Col. Stone's Life and Times of Red Jacket.]

Not far from this time, and with reference it is believed to this treaty, Thomas Morris says,—"Red Jacket was, I suppose, at that time about thirty or thirty-five years of age, of middle height, well formed, with an intelligent countenance, and a fine eye; and was in all respects a fine looking man. He was the most graceful public speaker I have ever known; his manner was most dignified and easy. He was fluent, and at times witty and sarcastic. He was quick and ready at reply. He pitted himself against Colonel Pickering, whom he sometimes foiled in argument. The colonel would sometimes become irritated and lose his temper; then Red Jacket would be delighted and show his dexterity in taking advantage of any unguarded assertion of the colonel's. He felt a conscious pride in the conviction that nature had done more for him, than for his antagonist."

"A year or two after this treaty, when Colonel Pickering from post master general, became secretary of war, I informed Red Jacket of his promotion. —'Ah!' said he,—'We began our public career about the same time; he knew how to read and write; I did not, and he has got ahead of me.—If I had known how to read and write I *should have got ahead of him*.'"

CHAPTER XI

Valley of the Genesee—Indian misgivings—Mill yard—Effort to obtain their land—Council at Big Tree—Coming of the Wadsworths—Indian villages—Refusal to sell—Discussion between Red Jacket and Thomas Morris—Breaking up of the Council.

The valley of the Genesee was a favorite resort of the Indian. His trail led along its banks and brought him at short intervals to Indian villages, or the head-quarters of Indian chiefs. Its flats were broad and beautiful, and were bordered on either side by hills that rose gradually to their summit, where they stretched out into extensive table lands. These hills, as we ascend the valley gradually become higher and higher, until we are brought into the vicinity of mountain elevations, where the scenery becomes very romantic, and the country much broken. The valley itself is almost of uniform width from its commencement, a few miles south of the city of Rochester, to the pleasant and thriving village of Mount Morris. Here these flats which are quite extensive and exceedingly rich and beautiful, appear to leave the river and follow its tributary, the Canaseraga, to a point about sixteen miles above; diminishing somewhat in width as they ascend, until they come near the present village of Dansville, where the hills again recede and forming a large basin, enclose it on the south, presenting the appearance of a magnificent amphitheater.

The Canaseraga is here joined by two streams, Stony Brook and Mill Creek, which flow down from the highlands beyond, over precipices, and through gorges deep and wild, where rugged cliffs defying all attempts at culture, rise abruptly at times, from one to three hundred feet on either side. The Indian's trail conducted him to these wilds, which still remain the most unchanged of all his ancient haunts. Here are solitudes seldom visited by man, where are treasured sublimities that enchain the mind, and inspire a feeling of devotion in the heart of the beholder. Here the Indian, undisturbed by other sights or sounds, may yet listen to the voice of the waterfall as it sounded in the ear of his fathers, or to the gentle murmur of the stream discoursing now, as it did to them, in passing hurriedly over its rocky bed. [Footnote: Who would ever suspect that a railroad would stride across any of these deep chasms? How presumptuous.]

Beyond this point the Canaseraga itself, as it flows from its source among the hills bordering on Pennsylvania, passes often through deep ravines, narrow defiles, and overhanging cliffs. The same is true also of the Genesee river above Mount Morris. Its course is marked by scenery rarely surpassed in sublimity and grandeur. [Footnote: The High Banks, as they are called, near Mt. Morris, and a similar formation, together with the falls, near Portage, have attracted the attention, and are often visited by the tourist.—J. N. H.]

The Indian as he followed his trail leading up along its banks, paused often to listen to the thunder of its waterfalls, or to watch its course while threading its way at the bottom of ravines, hundreds of feet beneath the jutting point where he was standing. The territory marked by this river was unsurpassed in the magnificence and beauty of its scenery, and in the variety and richness of its soil; and the Indian who lived for the most part in the open world, found here a home congenial to his spirit, and he loved it. The white man saw and loved it too. But he loved it not as the Indian, who looked upon it as already complete. The hills brought him venison, the valleys corn, and the streams on every side abounded in fish, the beautiful speckled trout, which fairly swarmed in all of these waters. What could he want more? He loved it as it was; just as it came from the forming hand of the Great Spirit.

The white man loved it for what he saw he could make of it; but how little he thought his making, would mar the desirableness and beauty of the Indian's home. He had already obtained of the Indian a title to all his land lying on the east side of this river. He had even been allowed to cross over to the west side, and look upon that generous *Mill Yard,* twelve miles square, as his own. A very extensive gift it is true, but as it was proposed to erect at the Genesee falls a saw mill, which was claimed to be a vastly benevolent institution, and would be useful to the Indians as well as whites, inasmuch as it would save the immense labor of splitting and hewing logs for plank, as they were going to make the water of the river split the logs and hew them at the same time; it was claimed that this surrender on the part of the Indians, would be but a just offset against the self-denial, great expense, and severe labor of the whites, in establishing so benign an institution as a *saw mill,* in these western wilds. This is one among many instances of the benevolence of the white man toward the Indian.

If the Genesee country was prized by the Indian, it was regarded with a wishful eye by the white man. And as he had obtained what was on the east side of the Genesee river, he was not content without a larger portion on the west. Already the tide of emigration had brought him to the utmost limit of his possessions, and he could hardly refrain from looking, with a wishful eye, upon the fertile fields lying beyond.

The Indian on the other hand, began to feel uneasy about having sold so much of his land. He regretted very much the permission he had given the white man to own one foot of ground, on the west side of the Genesee river. Natural boundaries with him weighed more than with the white man; and had the white man's possessions been confined strictly to the east side of the river, he would have felt better satisfied though it had cost him a larger area of ground. The white man's mode of running lines and of measuring land, he did not comprehend or appreciate. But when the line was made by a creek, river, or mountain, he understood it, and it harmonized better with his views of fitness, in dividing up the surface of this great earth. He was utterly unschooled in the art of computing by acres and roods. But the water's edge he had traversed with his light canoe, and with every point and islet on the lakes he was familiar. He had followed the rivers to where they came bubbling up from their rocky bed amid mountain elevations, and there was not a tributary stream or run, by whose side he had not rested, or by whose music he had not been charmed, keeping pace with it, as it went innocently busying and babbling along on its downward way. With any or all of these landmarks he was familiar, and when fixed upon as boundaries, he could readily recur to, and religiously keep them; for they had been made by the Great Spirit, and it was his life- study to know them.

Not satisfied with the large purchase already made, the white man contemplated still greater acquisitions of Indian land. Little did the red man suspect, while roaming unmolested over his native hills, that in civilized circles, the advantages and disadvantages of his cherished home were canvassed, and made the subject of negotiation and purchase. And it awakened his deepest surprise when assured, that without his knowledge or consent, his land had been sold. He was not aware that his ignorance of the value of his country, for the purposes of civilization, was made a subject of barter among his superiors in knowledge, and that men of enterprize were willing to pay for the privilege of making a bargain with him for his lands.

This right, as we have seen, was claimed by the government; Massachusetts holding the right of buying the Indian lands in Western New York. This right, under sanction of which the Phelps and Gorham purchase was made, was in part sold, as related in a preceding chapter. The pre-emptive right to the remainder was bought by Robert Morris in the spring of 1791. He re- sold soon after, to a company of gentlemen in Holland; pledging himself to survey the entire tract, and extinguish the Indian title.

Thirty-five thousand pounds sterling of the purchase money were retained, as a guaranty of his fulfilling these engagements.

It became an object therefore for Mr. Morris to obtain, at as early a period as practicable, a conference with the Indians, and their consent to sell this land. Owing to their extreme reluctance to part with any more land, he had not been able to persuade them to appoint a council for this purpose, and committed the further prosecution of this to his son Thomas. Hence the occasion given to notice the presence of Thomas Morris at the Indian councils, particularly that at Tioga Point. For several years he had been cultivating an acquaintance with the Indians, residing in their midst, attending their councils, and making himself generally agreeable; and by means of his own personal influence with the chiefs, and unwearied exertions he gained their permission to hold a council, which assembled at Big Tree, the present site of Genesee, in August, 1797.

This had already become the residence of the white man. James and William Wadsworth, from Durham, Conn., had emigrated hither as early as the year 1790. Under their auspices a new settlement had been commenced. On rising ground which commanded a fine view of the flats, stood their large block house. The same site has still its attractions, for what at a later day, was the old Wadsworth mansion.

The coming of the Wadsworths into this region, which was still in possession of the Indians, and their prominence in its subsequent history, would seem to justify a more extended notice.

In the spring of 1790, James Wadsworth, then a young man of twenty-two, was debating with himself the question of his future calling in life. He had graduated at Yale College in the fall of '87:—had spent the winter of '87 and '88, at Montreal, Canada, teaching school. He had no thought of teaching as his life-work, and what would he do next? was his earnest inquiry. Some one suggested that he should study medicine; but this did not suit him. As he had received a liberal education, it was further intimated that he should lead a professional life and become a lawyer, or a minister.

After duly considering the matter, choosing for this purpose the retirement of a neighboring wood, he returned the answer,—"I am not satisfied with either of these professions."

"What will you do, then?" was the inquiry. He replied, "I know God has made me for something, and I am trying to find out what that is."

With his mind thus unsettled, he determined to visit his uncle, Colonel Jeremiah Wadsworth, of Hartford. This uncle had pursued a sea-faring life, entering upon it at first for the benefit of his health, and following it afterward, from a love for the employment. From a sailor before the mast, he came to be mate, and captain, and at the breaking out of the Revolutionary war he had retired from the sea, and had settled at Hartford, Conn. He was appointed commissary of the Connecticut line, and subsequently had important trusts committed to his charge, by his own State, and also by the Congress at Philadelphia, having reference to the pay, clothing and subsistence of the Continental troops.

In the discharge of his official duties he had formed an acquaintance with Oliver Phelps; and after Mr. Phelps had secured an interest in the Genesee country, he represented its advantages to Colonel Wadsworth in such glowing colors, as led him to purchase a considerable tract of land in that region. Being a man of wealth and advanced in life, he had no thought of emigrating thither, but designed to provide for his interests by employing an agent.

As soon as James Wadsworth arrived at the house, he was met at the door by his uncle, who eagerly grasped his hand and exclaimed,—"James, I am glad you have come, you are the very man I have been wanting to see."

It was not long before they were deeply engaged in discussing the Genesee question, this becoming the chief topic of conversation during the visit. As the result, James purchased on advantageous terms a part of the tract at Big Tree, and became agent for the remaining lands, qualified by the condition that his brother William would consent to accompany him in the proposed emigration. [Footnote: Conversation of the author with Nehemiah Hubbard, Esq., of Middletown, Conn., and statements in Turner's History.]

The two brothers jointly entered upon the undertaking, and commenced preparations for their journey into this, at that time, far-off wilderness. An ox cart, and ox team, are in wide contrast with the conveniences of travel enjoyed at present. Yet with these, and two or three hired men, and a colored woman, a favorite slave belonging to the family, William set forth to encounter the vicissitudes and dangers involved in the enterprise. It was a slow and wearisome journey, most of the way rough, and some of the way requiring to be opened and prepared for travel.

James, with provisions and a small supply of household furniture, went by the sound, the Hudson, and the head of navigation on Canandaigua outlet. He arrived at Canandaigua three days in advance of his brother.

From this point their journey was comparatively easy. They pursued the route taken by Sullivan in '79, yet not without having frequently to cut a way for their team and cart. They arrived at their point of destination on the 10th of June, 1790.

Captain Horatio, and John H. Jones preceded the Wadsworths, and other families came into the region soon after. But the country was full of Indians. Their villages swarmed with life in every direction. Ken-de-wah or Big Tree, as principal chief was at the head of a numerous clan, located on the bluffs near by. Not far from them on the river was a village of the Tuscaroras. Two miles below was Oneida Town, a large village of Oneidas. Near the present site of West Avon was another principal village, whose chief was Ga-kwa-dia, or Hot Bread. Above was another large village called Little Beard's Town, occupying the present site of Cuylerville. Further on were Allen's Hill, Squaky Hill and Gardeau, the residence of the "White Woman." Her husband was principal chief of the clan at this point. Further on at Nunda, was another village, its principal chiefs were Elk Hunter and Green Coat. Still higher up on the river at Caneadea, was another considerable village, whose chief was John Hudson. [Footnote: It was here the author's grandfather, as an Indian prisoner, had to run the gauntlet in the spring of 1782.

The author remembers Hudson very well. Often visiting his grandfather's house in Angelica, N. Y. When a boy he often sat on Hudson's knee, whom he regarded as a very pleasant, kind Indian.]

These villages were mostly in the vicinity of Big Tree, a region which at that time was not without its charms, and has since been regarded as possessing attractions in soil and scenery, unsurpassed by any in the State.

It was here the council, solicited by Thomas Morris, assembled.

The unfinished block house of the Wadsworths was engaged for the accommodation of those particularly interested with Mr. Morris in conducting the council; and a large tent covered with boughs, and prepared with rows of seats, and a platform, furnished a place suited to their deliberations.

The United States, though not directly concerned as a party in this council, were interested in the welfare of the Indians, and appointed a commissioner to watch over their rights, and see that no injustice took place. Massachusetts reserved this right in the sale of her pre-emptive title. Accordingly Colonel Wadsworth of Connecticut, appeared as commissioner on the part of the United States, and General Wm. Shepard in behalf of the commonwealth of Massachusetts. William Bayard of New York represented the interests of the Holland company, and Mr. Morris, appeared through his agents, Thomas Morris and Colonel Williamson. The engagements of Mr. Williamson calling him away, the responsibility of conducting the treaty devolved upon Thomas Morris.

A large number of Indians were present, brought together by the prospect of good cheer, no less than their interest in the object of their assembling.

The council being duly opened, the commissioners offered their credentials, and explained the reason of their appointment; after which Mr. Morris presented in a speech of some length, the object for which they had been convened. Representing the desire of his father to obtain by purchase a part, or all of their lands, and how much better it would be for them to dispose of all, except what were actually needed for settlement, and place the money at interest, than to retain in their possession uncultivated wastes, whose only value to them could be such as were derived from the chase; and that this advantage would not be lost, for they could still use it for hunting, the same as before. He concluded by offering them the sum of one hundred thousand dollars, for the entire tract that remained to them in the State, allowing them such reservations as might be needed for actual use.

The Indians after deliberating for a time returned an unfavorable answer; saying "they did not wish to part with any more of their land."

Mr. Morris replied, urging them to reconsider the case, that they ought not to decide hastily, setting before them in various ways the favorable terms he had proposed.

After deliberating once more, they returned the answer they had already given.

Meetings and speeches thereupon succeeded; Farmer's Brother, Cornplanter, Little Billy, Little Beard, and Red Jacket, taking part in the discussion, the chief burden of which fell upon the latter.

When Mr. Morris urged upon their attention the liberal sum he had proposed to pay for their lands:—

Red Jacket replied,—"We are not yet convinced that it is best for us to dispose of them at any price."

"But," said Mr. Morris, "what value can they be to you as they now are, any further than the consciousness that you own them?"

"Yes," said Red Jacket, *but this knowledge is everything to us. It raises us in our own estimation. It creates in our bosoms a proud feeling which elevates us as a nation*. Observe the difference between the estimation in which a Seneca and an Oneida are held. We are courted, while the Oneidas are considered as a degraded people, fit only to make *brooms* and *baskets*. *Why this difference*? It is because the Senecas are known to be the proprietors of a *broad domain*, while the Oneidas are *cooped up in a narrow space*."

"Ah," said Mr. Morris, "you presume too much in regard to the consequence of your nation. It is far from being as great as you seem to suppose; and in proof of this let me refer you to the manner in which your deputation to the Miamis was received in 1793. Though large and composed of many of the first men of your nation, it had but little influence."

"Very true," replied Red Jacket, *"and why?* It was because we were in *bad company. We went with the pale faces.* Had we gone alone, we should have been treated with the *dignity which belongs to the Senecas throughout the world."*

While Red Jacket was still standing some one interposed the remark,—*"he's a coward."* Turning round with a look of contempt, and in tone and manner expressing the deepest sarcasm, he said,—"YES, I AM A COWARD." And then waving his hand over the broad and beautiful lands that were spread out before them, added: *"assure me that you can create lands like these, which the Great Spirit has made for us his red children, so that you can give us lands like them in return, and I will be brave:* UNTIL THEN, I AM A COWARD,—I DARE NOT SELL THESE LANDS."[Footnote: Conversation of the author with Wm. Jones.]

The commissioners together with the agent of the Holland company, who had been looking on and anxiously observing the proceedings for about two weeks, began to regard the undertaking as hopeless, and urged Mr. Morris to use more decisive means with them, and bring them to terms one way or the other.

Though contrary to Mr. Morris' convictions from his knowledge of the Indian character, as to its being the best method to pursue, he yielded to their solicitations; and when the Indians presented him the offer of a single township on the line of Pennsylvania, at one dollar an acre, Red Jacket assuring him that he could sell this at a sufficient advance, to pay for the trouble and expense of the treaty, he told them if that was all they could offer, they might return to their homes, for the sooner their conference was ended the better.

Red Jacket thereupon sprang to his feet and said, "You have now come to the point to which I wished to bring you. You told us in your first address, that even in the event of our not agreeing to sell our lands, we would part friends. Here then is my hand." Mr. Morris taking his hand, he then added; "I now cover up the council fire."

CHAPTER XII

Interview between Farmer's Brother and Thomas Morris—Mr. Morris addresses the women—Distributes presents—Negotiations continued—Treaty concluded with the women and warriors—Manner of payment—Inquiries about a Bank— Their reservations—White woman—Young King's dissent—Charge of insincerity.

The Indians appeared to regard the breaking up of the council at Big Tree, with great satisfaction. Their joy was unbounded; they made the forest ring with their wild yells, inveighing loudly and insultingly against Mr. Morris, and the commissioners, and assuming such menacing attitudes, as fairly to intimidate those unaccustomed to their rude manners.

To all present but Mr. Morris, the prospect of accomplishing any thing after this seemed utterly hopeless, and it was with some difficulty the commissioners were persuaded to remain, for the purpose of giving him the opportunity of another trial. Yet his hopes of success were so sanguine, as to induce them to tarry a short time longer.

The day after the breaking up of the council, Farmer's Brother called on Mr. Morris, expressed his regret at what had transpired, and the hope that it might not destroy the interest he had manifested for his nation. "Certainly not," said Mr. Morris,—"you had a right to refuse to sell your lands;" but he added, the treatment he had received from his people at the close of the council, especially in allowing a drunken warrior to menace and insult him; while they were yelling in approbation of his conduct, was uncalled for, and ungenerous. He had not deserved this from them. They had for years had food at his house in Canandaigua, and liquor as much as was for their good, and whenever any of them had been at Philadelphia, his father had treated them with equal kindness and hospitality.

Farmer's Brother acknowledged that all this was true, and regretted that the council fire had been extinguished so hastily, or they might have had a meeting, to smooth over these difficulties.

"Yes,"—said Mr. Morris, "and here is another ground of complaint. Red Jacket assumed the right of covering up the fire. This did not belong to him. For according to your custom, he only who kindles the council fire, has a right to cover it up."

"That is so,"—said Farmer's Brother.

"Then as I did not cover it up the council fire is still burning."

After thinking a moment he replied,—"*Yes:*"—and appeared to be pleased that it was so, and proposed to have the council convene again.

Mr. Morris signified his pleasure to delay a few days, to give him time to look over his accounts, pay for the provisions that had been consumed, collect his cattle that had not been slaughtered, and arrange other matters preparatory to his leaving the treaty ground.

He had become so well acquainted with Indian customs, that he had resolved upon another expedient, when his negotiations with the sachems had failed.

It is a rule among the Indians that their sachems shall have a right to transact whatever business belongs to their nation, whether relating to their lands, or anything else. But in transactions that concern their lands, if their course is not satisfactory to the women and warriors, they have a right to arrest the proceedings, and take the management into their own hands. The reason they assign for this practice is,—that the land belongs to the warriors, because they are the defense and strength of their nation, and to the women, because they are mothers of the warriors. In their polity therefore they recognize head or chief women, whose privilege it is to select a speaker to represent their views.

Mr. Morris determined as a last resort, to refer his case to the chief women and warriors. He accordingly sought and obtained such a meeting.

He made known to them his business, told them of the offer he had made their sachems, portrayed to them in glowing colors, the advantages they would receive from the annuity so large a sum would bring,—how it would furnish them with food and clothing, without any anxiety or toil on their part, and that they would thus be relieved of many hardships, which they were now compelled to endure.—That the sachems, who were unwilling to sell the land, always had enough to supply their wants.—That they could kill game, and feast on the meat, and go to the settlements and sell the skins, and buy them clothing. Hence they did not care to exchange their land for money, that would enable the women to obtain for themselves and children food and clothing, whereas they were now often compelled to go hungry and naked. By selling such a portion of their land as they had no use for, they would have the means of supplying their necessary wants, and of making themselves comfortable. He then displayed before them a large supply of beads, blankets, silver brooches, and various other ornaments, of which the natives were particularly fond, and said he had brought these with the design of making them presents, in the event of a successful treaty. But in as much as the women were not to blame for breaking off the negotiations, he was determined they should have the presents he had intended for them. He accordingly proceeded to distribute among sparkling eyes, and joyous hearts, the beauties and treasures, he had brought for them.

These gifts proved a most powerful addition to his argument, and were the means of giving a favorable turn to their counsels. For several days after this the chiefs, and women, and warriors, could be seen scattered about here and there in small parties, engaged in earnest conversation, which resulted in a renewal of their negotiations.

Mr. Morris was informed that their council fire was still burning, and that their business might proceed,—but instead of being carried on by the sachems, would be conducted by the women and warriors.

Cornplanter being the principal war-chief, appeared on this occasion in their behalf.

He said,—"They had seen with regret the misconduct of the sachems; that they thought also the action of Mr. Morris was too hasty; but still they were willing the negotiations should be renewed; and hoped they would be conducted with better temper on both sides."

Mr. Morris offered a few conciliatory remarks in reply; and Farmer's Brother, on the part of the sachems, represented these proceedings of the women and warriors, as in accordance with the customs of their nation.

The way being thus opened, the negotiations were readily carried forward to a successful termination.

They consented to sell their lands for the sum proposed, which was one hundred thousand dollars, leaving their reservations to be settled, as they could agree.

The simplicity of the Indian character was apparent, in the eagerness with which they desired to know about a *Bank:* the president having directed that the money they received for their lands, in case they were sold, should be invested for their benefit in stock of the United States Bank; in the name of the president, and his successors in office, as trustees of the Indians; they earnestly inquired,—*what is a Bank*?

Several attempts were made at explanation, when finally they came to understand, that the United States Bank, at Philadelphia, was a large place where their money would be planted, and where it would grow, like corn in the field.

As it was desirable also for them to understand, that the dividends from it might be greater some years than others, this was explained by referring to the idea of planting, as they could know from experience, that some years they would have from the same ground a better crop than others. Hence after this when speaking to Mr. Morris about their money, they would inquire *what kind of a crop they were going to have that year*?

Another point of interest with them, was to ascertain *how large a pile, the money they were to receive, would make*?

This was shown them by representing the number of kegs of a given size, it would take to hold, and the number of horses that would be necessary to draw it.

These questions being settled, the next point to be agreed upon, was the size of their reservations. Mr. Morris had stipulated, in case their demands were reasonable, no deduction would be made from the price they were to receive. But instead of moderate, very exhorbitant claims were presented, growing out of a degree of rivalry between different chiefs.

Their comparative importance would be graduated in a measure by the size of their domain, and the number of people they would thus be enabled to have about them; hence they were individually ambitious of not being out- done, in the size of their reservations.

Red Jacket put in a claim to about one-fourth of the entire tract purchased. Cornplanter desired about as much; and other chiefs were alike ambitious in securing extensive reservations; and they wished to have them marked out by natural boundaries, such as rivers, hills or the course of streams. To all of these demands Mr. Morris was obliged to give a stout and resolute denial, requiring them to fix upon a certain number of square miles, which, in the aggregate, should not be far from three hundred and fifty.

Here also arose difficulties about the size of their respective allotments, which they were unable to settle, so that Mr. Morris was obliged to assume the office of arbiter, and decide these for them, which he accomplished generally to their satisfaction.

In only one instance did he depart from his purpose of not allowing natural boundaries, in describing their reservations. It was in case of Mary Jemison, the White Woman, who lived on the Genesee river, some few miles above Mt. Morris. Her history is one of singular interest, and as belonging to this region, and connected with the circumstances under consideration, a brief notice of this remarkable woman, will not be out of place.

Hers is an instance of the entire change that may be wrought, in the taste and inclination, so that instead of a civilized, a person may prefer an uncultivated state of society. Though descended from the whites, she became so thoroughly Indian in her feelings and habits, that she was regarded as a curiosity, and called by way of distinction—the *"White Woman."*

She was born on the ocean, while her parents were emigrating from Ireland to this country, about the year 1742 or 3. Her father and mother soon after landing at Philadelphia, removed to a frontier settlement of Pennsylvania, lying on what was called Marsh Creek. During the war between the French and English, she was taken captive with her parents, by a party of Shawnee Indians. On the way, her father and mother were killed. The mother anticipating, from tokens she had observed, what would be their fate, advised her child not to attempt an escape from the Indians, as she most likely would be taken again, and treated worse. But as a course better adapted to promote her welfare, she was told to try and please her captors, adding as her parting counsel,—"don't forget, my daughter, the prayers I have taught you,—repeat them often; be a good child, and may God bless you."

After this, under various trials she went with the party, until they came to Fort Du Quesne. [Footnote: Afterwards called Fort Pitt, now the site of Pittsburg.] Here she was given to two Indian women, who were of the Seneca nation, and lived eighty miles below, on the Ohio river, at a place called She-nan-jee. With the usual ceremony observed by the Indians on such occasions, she was adopted into their family, and called De-ha-wa-mis. At length under kind treatment she began to feel as one of them. In time she was married to a young chief of the Delaware tribe, with whom she lived happily for several years in the Shawnee country. She became devotedly attached to her Indian husband, who treated her with marked tokens of affection.

After a time she welcomed with the joy of a young mother's heart, the appearance in her wigwam of a daughter, her first born. The bright morning of her domestic joy was soon overcast with sorrow; she is seen strewing over her little one's grave, the fallen leaves of autumn. She-nin-jee, her Indian husband once more became a father. Together they gladly embraced a son. Their lonely cabin after this was enlivened and cheered by his childish prattle; nothing now remained to interrupt the joy of the mother, but the absence of the father, whom the season of hunting, took far away from his cherished home. Yet with returning spring these toils are forgotten, as he is surrounded once more with the charms of the domestic fireside. But at length there came a spring whose joyful return, brought not the long wished for She-nin-jee, back to his lonely cabin. Many an evening fire blazed brightly to bid him welcome, yet he did not come. Choice venison had been dried and laid up for him, new skins had been prepared and spread for his couch, and many a silent hour whiled away with thoughts of the absent one, but he came not. His returning comrades brought back the sad news of his death. De-ha-wa-mis mourned long and deeply for the pride of her Indian wigwam. Her own kindred could not have extended to her more genuine sympathy, than did her new relatives by adoption. They kindly offered to take her back, if she desired to go, to her former friends among the whites, or if she chose to remain among them, they promised to give her a home of her own.

A part of her Indian relatives lived in the valley of the Genesee, and this was the occasion of her removal there, from her home on the Ohio. A few years intervened, and she again became the wife of an Indian, the distinguished Seneca warrior Hio-ka-too. She resided with him until his death, at Gardeau, the place where she was living, at the time of her appearance at this treaty. The chiefs desired for her a special reservation. To this Mr. Morris readily assented, in case she would specify a certain number of acres.

She said to him,—"I do not know any thing about acres, but I have some improved places;" pointing them out on the ground; "here a patch of potatoes, there, a few beans, and another still, where there's a little corn." She wished these might be embraced in her reservation, at the same time giving boundaries, which she thought would include them.

Mr. Morris owing to the lateness of the hour, and the impatience of the commissioners, gave his consent to the boundary named, supposing it might include a hundred or a hundred and fifty acres. But much to his surprise, the tract when surveyed, was found to contain not less than seventeen thousand, nine hundred and twenty-seven acres. [Footnote: Indian Treaties, p. 39. This reservation has been variously represented to contain, four thousand, and by others a larger number of acres. Col. Stone makes it thirty thousand. The amount given in the text is that obtained by actual survey of the boundaries in question. They are as follows: "Beginning at the mouth of Steep Hill Creek, thence due east until it strikes the Old Path, thence south until a due west line will intersect with certain steep rocks on the west side of the Genesee river, thence extending due west, due north, and due east, until it strikes the first mentioned bound, enclosing as much land on the west side, as on the east side of the river."

The survey by Augustus Porter, surveyor, gives it 17,927.]

Mary Jemison, the White Woman, had thus secured to her, according to the pledge of the Indians to give her a home, a princely domain, where for years after in primitive simplicity, she planted her beans, potatoes and corn, and maintained, as in former years, the usages of her Indian life.

The most of this tract she afterward sold to John Grey and Henry B. Gibson of Canandaigua; a deed for which was executed bearing date of September 3d, 1823.

She retained for her own use twelve hundred and eighty acres, and received for the balance, the sum of four thousand two hundred and eighty-six dollars, or an annuity of three hundred dollars forever.

The Senecas became gradually dispossessed of their lands in the valley of the Genesee, and in the year 1825, removed to their reservation at Buffalo. At the time of their removal, the White Woman refused to part with the residue of her land, and continued to reside at the place, where she had passed the greater part of her long life, and which was now endeared to her by many associations in the past.

But here she soon found herself surrounded by another race, and as time advanced, she longed to be among the people she had chosen for her kindred, and disposing of her possessions in the Genesee valley, removed to Buffalo in 1831.

She had now upon her the infirmities of age. Long had the parting injunction of her christian mother passed from memory. The religion as well as habits of the Indian, had become hers. Ninety summers had passed over her head. The missionary had visited her, and had been assured that her faith had long been in accord with that of the red man, and she had no desire to change her religious views.

But ere her last hour came a voice reached her from the distant past. It awakened memories long forgotten. She sent for the missionary. He came and stood by her. She was almost withered away. Her small, shrivelled, finely wrinkled face, silvery hair, toothless mouth, the nose almost touching her chin, and her thin, wasted form, indicated the presence of second childhood. The memory of that long lost mother rushed back upon her mind. She cried out in anguish, as well as sincerity of heart, "Oh, God! have mercy upon me!" The prayer of her childhood returned; she instinctively began to say.—"*Our Father which art in heaven.*"

As a child she received the instructions of the missionary, and before departing this life, her soul was lighted up with a cheering hope, based upon a reception of the clear and living truths of Christianity.

No one had sought to disturb the serenity of her advanced life, by intruding upon her the idea that she was a sinner. How came she to be thus exercised? The lessons given in childhood, availed more than sermons, and impressions were then made, which though apparently effaced, still remained to be quickened into life, and bring forth fruit, which cheered the closing days of her singularly eventful history.

With the settlement of the White Woman's reservation, Mr. Morris regarded the business of the treaty, as about concluded. Yet a new obstacle was presented by the arrival of Young King, a descendant of "Old Smoke," a renowned chief, held in great veneration among the Senecas. None had ever attained a greater degree of power, or swayed a more commanding influence. The son though not possessing the high endowments of the father, yet when he chose to exert it, commanded an extensive hereditary influence, which carried with it great weight. Having been informed of the proceedings of the council thus far, he expressed his disapproval.

Cornplanter and Farmer's Brother informed Mr. Morris that the treaty could not be completed contrary to the wishes of Young King; that however unreasonable it might appear to him, for one man to defeat the will of the entire nation, it was a power he received from his birth, and one of which he could not be deprived. Yet after much persuasion, Young King, though not reconciled to the idea of selling their lands, acquiesced; saying—"he would not stand out against the wishes of his nation."

The signing of the treaty yet remained; and Red Jacket according to the testimony of Mr. Morris, though he had strenuously resisted the sale, desired nevertheless to have his name appear among the chiefs of his nation, whose signatures were appended to the deed executed on the 15th of September, 1797, conveying to Robert Morris of Philadelphia, the title to all their lands west of the Genesee river, not included in their reservations, or previously sold.

From this fact the inference has been derived, that the orator was *insincere* in his opposition to the sales made of his people's lands. His sincerity though questioned now, was never after this a matter of doubt. If he had been insincere before, the effect of this sale on the destiny of his people, imposed upon him considerations of so grave a nature, as to render the idea of his indifference extremely improbable, and no one after this ever thought of imputing to him such a motive. Yet in all the sales the Senecas made of their land, subsequent to this period, Red Jacket's name, however much he may have resisted the act, was appended to the deed or instrument of conveyance. The reason he assigned for this, was his desire to have his name go, whether for better or worse, with the destinies of his people. Having exerted all his energies to prevent the sale of their lands, he felt that his duty had been discharged. And when his people decided against him, he regarded the responsibility of the transaction as resting on those who had effected it, and whether he gave or witheld his name, it would have no influence in determining the result.

He may have had some pride also in having his name appended to a document, which he knew the white people regarded, as of much importance, and were very careful to preserve.

It is related of him as having transpired at a later period, when Mr. Greig of Canandaigua, acting for the Ogden Company, was holding a council with a view to purchase some of the smaller Indian reservations, lying along the Genesee river, he was opposed step by step, by the persistent efforts of Red Jacket. Yet notwithstanding the opposition, Mr. Greig was successful in securing the extinguishment of their title, to about eighty thousand acres of their land. When the time came for signing the deed, Mr. Greig said to Red Jacket,—"*As you have been opposed to the sale of the land, you need not have your name attached to the deed.*" But he would hear to nothing of the kind, and insisted upon signing it, seeming to take pride in having his name appended to the paper. [Footnote: Conversation of the author with the Hon. John Greig of Canandaigua. Some years ago a story illustrating the eloquence of Red Jacket went the rounds of the papers, in which Mr. Greig was represented as arguing a case in opposition to and as being defeated by Red Jacket. Not happening to see it at the time, the author sought for a copy, but learning that its principal statements were fictitious, he relinquished the undertaking. Mr. Greig never argued the case as represented, but took down a speech from the interpreter which he read to the orator, who was much pleased with its correctness and bestowed on him an Indian name, signifying—"*a ready writer.*"]

CHAPTER XIII

Council at Canawangus—Interesting reminiscence of Red Jacket—Address of
Farmer's Brother—Jasper Parrish—Horatio Jones—Red Jacket's visit at Hartford, Conn.

A council of the Iroquois was held at Ca-na-wau-gus, near West Avon, in the autumn of 1798. Connected with it is a reminiscence of Red Jacket of much interest, as an item of history, and it serves well to illustrate the orator's mental habit.

His conduct was such on this occasion, as to excite the observation as well as curiosity of Captain Parrish, who related the occurrence.

For the first few days of the council, he uttered not a word. He appeared to be in deep thought, and was exceedingly reserved.—The expression of his countenance was severe, and there was much *hauteur* in his manner. He ate scarcely anything, and his appearance was so remarkable, as to excite the wonder of all present. At length on the third or fourth day of the council, he arose with great dignity, and solemnity of air, and commenced speaking. His exordium was for the most part a beautiful and highly wrought enconium on the character and history of the Indians; particularly of his own people, in the past. They were taken back, as by a magic spell, to primitive times. The days of their renown, when the name and glory of their nation, were the admiration of the world. When from the rising to the setting sun, there was no power to stand before them, or hinder the victorious march of their warriors through the land. As they glided over the waters of river or lake, as they ascended the mountain, or passed through the valley, they could feel that their dominion was wide, and undisputed. Every deer that bounded through the forest, every bird that winged the air, and the fish in all the waters, were theirs, and they were happy. Such was the glowing picture he drew, they did not realize the present, from the engrossing theme of the past.

He next proceeded to sketch their history, as affected by the coming of the white man among them. The friendly relations, that marked their early intercourse. Their small beginnings, and the imperceptible manner of their increase. How they began to line the eastern shores,—plant themselves upon the borders of their rivers, and gather into neighborhoods, and towns, and cities. How these new and wonderful things engaged the attention of the Indians, and kept them spell-bound, so that they were insensible to what had been going on till the whites were firmly planted, like a tree that has taken deep root, and sends its branches out over the land.

He next drew their attention to a time when the signs of a great tempest began to appear. When the clouds began to overspread the heavens, when the lightning flashed, and the thunders rolled, and the land was shaken by their power. A mighty whirlwind came sweeping through the land, the tall trees of the forest were uprooted, the branches torn off and sent flying through the air. So has our nation he said been uprooted,—the strong men torn from us, and scattered, and laid low. Thus he went on recounting as few could, the circumstances of their history, and as he advanced, his expressions matured in their intensity, his thoughts appeared to be winged, and came glowing, as if from some furnace in nature, where all her materials are wrought under intense heat, and sent forth in forms of highest brilliancy, and beauty. His hearers were amid the heavings of the earthquake,—the blackness of the storms,—the wild and irresistible sweep of the tornado. The heavens, the earth, the elements, seemed to be careering under the rapid and startling flights of his fancy.

He next adverted with much feeling, and with evident sadness, to the transactions of the past year, by which they had become dispossessed of the largest part of their ancient inheritance: and then he drew, with a prophetic hand, a picture of their probable future, that brought sorrow to their hearts, and tears to their eyes.—He closed his harangue by pronouncing a most withering phillipic against the whites.—The effect of his speech was wonderful.—Mr. Parrish declared that it exceeded, in its brilliancy and force, all his former utterances, of which he had any knowledge; and he never heard from him afterward, anything that could compare with it. His auditors were mainly those of his own people. His flow of thought was not interrupted by the slow, and embarrassing process of interpretation. The full grief of his heart, in view of the transactions of the previous year, was poured forth, and came like the irresistible sweep of a whirlwind. [Footnote: Conversation of the author with Samuel J. Mills, Esq., formerly of Mt. Morris, N. Y., later of Nevada, Iowa. Mr. Mills heard Mr. Parrish give this description of Red Jacket and of his speech, while sitting at one time on the porch of one of the hotels at Avon Springs. Mr. Parrish pointed out the ground occupied by the Indians, when this speech was delivered. It was only a little distance from the porch where they were sitting.]

It was some little time after the delivery of this speech, before the minds of the Indians were sufficiently composed to attend to the main business of their council, which was presented in a speech by Farmer's Brother, and embodied in an address to the Legislature of New York, thus: "The Sachems, Chiefs and Warriors of the Seneca Nation, to the Sachems, and Chiefs assembled about the great Council Fire of the State of New York:

"BROTHERS: As you are once more assembled in council for the purpose of doing honor to yourselves and justice to your country, we, your brothers, the Sachems, Chiefs and Warriors of the Seneca Nation, request you to open your ears, and give attention to our voice and wishes.

"You will recollect the late contest between you and your father, the great king of England. This contest threw the inhabitants of this whole island into a great tumult and commotion, like a raging whirlwind, which tears up the trees, and tosses to and fro the leaves, so that no one knows whence they come, or where they will fall.

"BROTHERS: This whirlwind was so directed by the Great Spirit above, as to throw into our arms two of your infant children, Jasper Parrish, and Horatio Jones. We adopted them into our families, and made them our children. We loved them and nourished them. They lived with us many years. At length the Great Spirit spoke to the whirlwind, and it was still. A clear and uninterrupted sky appeared. The path of peace was opened, and the chain of friendship was once more made bright. Then these, our adopted children, left us, to seek their relations. We wished them to return among us, and promised if they would return, and live in our country, to give each of them a seat of land for them, and their children to sit down upon.

"BROTHERS: They have returned, and have for several years past been serviceable to us as interpreters. We still feel our hearts beat in affection for them, and now wish to fulfil the promise we made them, and to reward them for their services. We have therefore made up our minds to give them a seat of two square miles of land lying on the outlet of Lake Erie, about three miles below Black Rock, beginning at the mouth of a creek known by the name of Scoy-gu-quoy-des Creek, running one mile from the river Niagara, up said creek, thence northerly as the river runs two miles, thence westerly one mile to the river, thence up the river as the river runs to the place of beginning, so as to contain two square miles.

"BROTHERS: We have now made known to you our minds; we expect and earnestly request that you will permit our friends to receive this our gift, and will make the same good to them, according to the laws and customs of your nation."

By the laws of the State, no sale or transfers of Indian lands could be made to private individuals, without permission from the government.

Hence the address embodying the request as presented above, which was complied with, and the land secured as desired by the Indians.

The above is certainly an able document, and has been justly admired for its originality, and the boldness of its figures. It is in keeping with the high order of mind, that has marked the history of the Six Nations. One expression in it has been pointed out, as an instance of the truly sublime: "THE GREAT SPIRIT SPOKE TO THE WHIRLWIND, AND IT WAS STILL."

We may observe here that in tracing the history of the Iroquois, the instances are not rare of a true nobility of character. Their confidence and esteem once secured, no slight cause would interrupt, none appreciated more highly the offices of kindness,—and none would go further in making a generous return for favors rendered.

Jasper Parrish and Horatio Jones were favorite interpreters of Red Jacket, and as they passed no inconsiderable part of their lives among the Indians, a further notice of their history is desirable.

The early life of Captain Jasper Parrish was marked by scenes alike trying and eventful. He was a native of Connecticut, from which State his family removed to the waters of the Delaware, in the state of Pennsylvania. In 1778, when but eleven years old, he accompanied his father on a short expedition, to remove a family of backwoodsmen, to a less exposed part of the settlement. On their way they were attacked by a small party of Indians, and made captives. The father was taken to Niagara, and after a captivity of two years, was exchanged and enabled to return to his own family.

The son was claimed by a war-chief, who treated him kindly, and after a time took him to the waters of the Chemung. On entering an Indian village, the war-party which accompanied them, sounded the war-whoop, and it was answered by the Indians and Indian boys who came out to meet them. They pulled the young prisoner from the horse he was riding, scourged him with whips, and beat him with the handles of their tomahawks, one of the forms of their gauntlet, until his master humanely rescued him. He was after this sold to a family of Delawares, and taken to reside with them on the Delaware river, where he suffered much from want of proper clothing, and from scanty fare. To inure him to cold, the Indians compelled him almost daily, to strip and plunge into the icy waters of the river.

He was with the Indians when General Sullivan invaded their country, and witnessed their retreat, after the battle at Newtown, until they found protection from the guns of the British, at Fort Niagara. Here they subsisted during the winter by rations from the garrison, and to induce them to return again to their villages, on the Genesee river, the officers pledged them an increased bounty for American scalps.

On one occasion, while with the Delaware family at Niagara, he came near being a victim of the British bounty for scalps. Left alone with some Indians, who were having a carousal, he overheard a proposal to kill the young Yankee, and take his scalp to the fort, and sell it for rum. In a few moments one of them took a large brand from the fire and hurled at him, but being on the alert he dodged it, and made his escape. The Indians pursued, but it was dark and they did not find him.

From the Delaware family, he was sold to an Indian of the Mohawk tribe, called Captain David Hill. At a council of the British and Indians, he was afterwards adopted with much ceremony, into the family of Captain Hill, as his own son. He resided with him at the Mohawk settlement near the present village of Lewiston, till the close of the war, and being surrendered in accordance with the stipulations of the treaty at Fort Stanwix in 1784, he returned once more to his own father's house.

It was with some effort he recovered again the use of his own native tongue. During his captivity he had acquired and could speak fluently, the language of five different tribes, and his qualifications as an interpreter, together with his known faithfulness and integrity, coming to the knowledge of our government, he received an appointment in the Indian service, and during the greater part of his subsequent life, was actively employed in business relating to the welfare of the Indians. He died at his residence in Canandaigua, July 12th, 1836, in the sixty-ninth year of his age.

Captain Horatio Jones, was a native of Chester county, Pennsylvania. At the age of sixteen he enlisted as a volunteer, in a company commanded by Captain John Boyd. It was when the Indians, led by the notorious Butler, Brant, and Nellis, were committing their depredations and massacres among the settlers of the frontier, sparing neither age nor sex, from the tomahawk and scalping knife. With the ardor of youth he engaged in the active employments of a soldier, and accompanied Captain Boyd on several important and dangerous expeditions, in which himself and commander had the good fortune to escape unhurt.

At length in the spring of 1781, while Captain Boyd and his men, numbering thirty-two, were in pursuit of Nellis, they were surprised by a large party of Indians, who killed about half their number, and of the rest took eight prisoners, Jones and his commanding officer being among the number. The Indians conducted them to their towns on the Genesee river, where they had to run the gauntlet, and having passed with safety through this trying ordeal, they next came near losing their lives in a savage frolic. The warriors, on returning from their excursion, gave themselves up to drinking and merriment. Partaking freely of the intoxicating bowl, they soon became much excited, and the ferocity, which a time of war engenders, was thoroughly aroused among them. One of the prisoners they killed, and severing his head from the body, carried it about the camp, on the end of a pole, with wild shouts and frantic yells.

They next meditated the death of Boyd and Jones, and while discussing the manner in which they would have them suffer, a few squaws conveyed them away and hid them. Jones was subsequently adopted into an Indian family, became familiar with their customs and language, and after the declaration of peace, was appointed by President Washington as Indian interpreter, the duties of which office he discharged with fidelity, until within a year or two of his death.

Mr. Jones was about the ordinary stature, firmly built, and qualified by nature for duties requiring activity and endurance. Possessing uncommon mental vigor, and quick perception, he was enabled to form a just estimate of character, and determine with readiness the springs of human action. His bravery, physical power, energy and decision of character, gave him great command over the Indians with whom he was associated, and having their entire confidence, he was enabled to render the government invaluable service in her treaties with the northern and north-western tribes. He was a favorite interpreter of Red Jacket, and his style is said to have been energetic, graphic, and chaste. He died at his residence near Genesee, on the 18th of August, 1836.

It was not far from the time of this council at Canawangus that Red Jacket visited Hartford, Conn.

In the adjustment of the land difficulties between the states of Connecticut and Pennsylvania, owing to the indefinite terms of their original charters, Connecticut obtained, as we have seen, a title to that part of Ohio, called Western Reserve. The Senecas laying claim to this, on the ground of conquest, negotiations were entered into with them for the extinguishment of their title. This was the occasion of the orator's visit, concerning which there is but a very brief record. His appearance, however, has been spoken of in terms of high commendation, and a single passage only of the speech he made on that occasion has been preserved.

"We stand," —said he, when representing the condition of his people,— "a small island in the bosom of the great waters. We are encircled,—we are encompassed. The evil spirit rides upon the blast, and the waters are disturbed. They rise, they press upon us, and the waves once settled over us, we disappear forever. Who then lives to mourn us? None. What marks our extermination? Nothing. We are mingled with the common elements."

The entire speech was listened to with feelings of profound admiration, and his action elicited praise for its dignity and grace. He entered the august assemblage, before which he was called to appear, with a step measured, firm and dignified,—a countenance erect, bold and discursive,— without manifesting surprise, fear or curiosity; and his effort sustaining fully his high reputation as an orator, made the occasion one of great interest, to those whom it had been the means of bringing together, or who had been attracted by curiosity, to see one whose fame had reached the land of steady habits. [Footnote: Col. Stone, from collections by J. W. Moulton.]

CHAPTER XIV

Cornplanter in disrepute—Effort to regain his standing—Red Jacket charged with witchcraft—His defense—Further notices of Cornplanter—Early recollections—At the defeat of General Braddock in 1755—With the English in the war of the Revolution—Takes his father a prisoner—His address—Releases him—Address to the Governor of Pennsylvania—Visit of President Alden—Close of his life.

Not long after the large sale of their domain to Robert Morris, which had been negotiated at Big Tree, the Senecas began to realize that they had committed a great mistake. The broad lands, mountain, hill, and valley, over which they had roamed, the springs and streams of water by whose side they had been wont to encamp, and above all the graves of their sires, where affection's altar had been hallowed by their sighs and tears, these were still in view, but they appeared not as in days gone by, to wear for them the smiles of old and long tried friends. They seemed to present a look and utter a voice of reproach, as though chiding them for having broken in upon the harmony of those time honored arrangements, which had bound them together, and the thought of this filled their minds with anxiety and grief. Had they been aware of the sorrow they would experience in looking upon these lands, as no longer their own, their consent to part with them would not so readily have been given.

The reverse which thereupon took place in their minds, fell heavily on those who had taken the most active part of the business of selling their country. Cornplanter, having borne a prominent part in these proceedings, fell deeply under the displeasure of his people. Their displeasure was so marked as to lead him to cast about for some means of relief. Aware of the credulity and superstition of his people, he resolved to avail himself of these characteristics of his nation, to accomplish the end he had in view.

For this purpose he was in consultation with his brother Ga-ne-o-di-yo, who on one occasion terminated a scene of great dissipation, by the announcement that he had been delegated by the Great Spirit, with a new revelation, and with supernatural gifts. A severe illness became the occasion during which he made a visit to the unseen world, where visions and revelations of a most extraordinary nature, had been made known to him. The happiness of the good, and the tortures of the wicked, had thus become matters of personal observation. The announcement of these, in language and gesture indicating his assurance of their reality, gained for him credence among the people, as well as chiefs of his nation, and he was received as a prophet.

His earliest attempts were successful in accomplishing a desirable reform, especially among the Onondagas, the most profligate of the Six Nations, from the degrading vice of intemperance. His influence in this direction was salutary, and had he confined his efforts to the recovery of his people from drunkenness, his mission would truly have been one of mercy, and his career might have terminated with the highest usefulness and honor.

But sympathizing with Cornplanter, his brother, he conceived the idea of instituting against their enemies, the charge of *witchcraft*. In this the Indians generally believed, and a charge of this nature, coming from such a source, was a very grave matter. Through the instrumentality of Congress selected by himself, the sentence of death was procured against certain "familiars of Satan," and this sentence would have been executed, had there been no interference, from the knowledge of it coming to the whites, living in the vicinity.

In no way discouraged, but rather emboldened by their success, they proceeded so far as to bring such a charge against Red Jacket himself, who was thus publicly denounced, at a great council held at Buffalo Creek, and put upon trial.

A degree of rivalry had hitherto existed between Cornplanter and Red Jacket, and as the former descended in the estimation of his people, for the part he had taken in the sale of their lands, the latter rose for the same reason, so that the highest aim of Cornplanter was reached, when he could, by this means, affect materially the character, and influence of his distinguished rival.

The orator was thus placed in circumstances the most critical and trying, of any that had hitherto met him in life. He perceived at a glance, that his entire history in the future, would depend on the decision that would then and there take place. He might be doomed, if his life were spared, and this was not altogether certain, to be the victim of surmises and superstitions, that would be annoying, if they did not prove to be utterly destructive of his happiness. He accordingly summoned himself for an effort as great, as his position was dangerous.

He conducted the trial in his own defense. In this he exhibited the exceeding wariness, which was ever a prominent characteristic of his nature.

The slightest circumstance affecting the character, or bearing suspiciously upon his adversary was not overlooked, and his history was scanned with the searching scrutiny of a mind, that seemed to grasp intuitively, the secret springs, which had influenced his conduct. One by one the professions that had formed his garb of sanctity, were exposed to the burning power of his keen satire, and step by step he advanced to a point, where, from the full assurance he had established this conviction in the minds of his people, he pronounced him AN IMPOSTER,—A CHEAT. [Footnote: Conversation of the author with Wm. Jones, a chief among the Senecas, and a son-in-law of Red Jacket.]

His speech riveted the attention of his hearers for nearly three hours. He prevailed. "THE IRON BROW OF SUPERSTITION RELENTED UNDER THE MAGIC POWER OF HIS ELOQUENCE."—The Indians divided and a majority appeared in his favor.

"Perhaps,"—says the distinguished author just quoted,—"the annals of history cannot furnish a more conspicuous instance of the triumph and power of oratory, in a barbarous nation, devoted to superstition, and looking up to the accuser as a delegated minister of the Almighty." [Footnote: Governor Clinton's Historical Discourse.]

The victory which Red Jacket thus achieved recoiled heavily on Cornplanter, and gave him a blow, from which he never afterward fully recovered. He retired to his reservation, on the waters of the Alleghany river, within the boundaries of Pennsylvania, where he devoted himself, during the remainder of his long life, to the elevation and improvement of his people. He did not, after the example of his great rival Red Jacket, spurn the improvements of civilization, but engaged in agriculture after the example of the whites, and welcomed to his abode the teachers of christianity, and himself openly avowed his belief in its doctrines.

Cornplanter was a native of Ca-na-wan-gus, on the Genesee river, a half breed, the son of an Indian trader, from the valley of the Mohawk, a white man named John O'Bail. Of his early life little is known further than he himself intimated, in a letter written long afterward, to the governor of Pennsylvania:—In which he said,—"When I was a child I played with the butterfly, the grasshopper, and the frogs; and as I grew up, I began to pay some attention, and play with the Indian boys in the neighborhood; and they took notice of my skin, being a different color from theirs and spoke about it. I inquired of my mother the cause, and she told me that my father was a resident in Albany. I still ate my victuals out of a bark dish. I grew up to be a young man, and married me a wife, and I had no kettle or gun. I then knew where my father lived, and went to see him, and found he was a white man, and spoke the English language. He gave me victuals, while I was at his house, but when I started to return home, he gave me no provision to eat on the way. He gave me neither kettle or gun."

He was with his people when they fought in alliance with the French in the year 1755. The principal part of the force which met and defeated the English under General Braddock was Indian, and it was through their prowess mainly, if not entirely, that the victory was gained.

What part Cornplanter took in that engagement is not known, but in the war of the Revolution, he was a war-chief, and ranked high in the estimation of his people.

In a speech addressed to President Washington in 1790, he related the manner in which the Indians came to be in alliance with the English.

"Many nations inhabited this country; but they had no wisdom, therefore they warred together. The Six Nations were powerful and compelled them to peace; the lands to a great extent were given up to them; the French came among us and built Niagara; they became our fathers and took care of us. Sir William Johnson came and took that fort from the French; he became our father and promised to take care of us, and did so until you were too strong for his king.

"When you kindled your thirteen fires separately, the wise men that assembled at them told us that you were all brothers, the children of one great father, who regarded the red people also as his children. They called us brothers, and invited us to his protection; they told us that he resided beyond the great water, where the sun first rises; that he was a king whose power no people could resist, and that his goodness was as bright as that sun. What they said went to our hearts; we accepted the invitation, and promised to obey him. What the Seneca Nation promise, they faithfully perform; and when you refused obedience to that king, he commanded us to assist his beloved men, in making you sober. In obeying him we did no more than yourselves had led us to promise. The men that claimed this promise told us that you were children, and had no guns; that when they had shaken you, you would submit. We hearkened to them and were deceived."

As a leader he was very active and brave, and as a partisan of the English, bore a prominent part in all of the principal engagements, in which the Indians were concerned during that war. He was on the war-path with Brant during the campaign of General Sullivan against the Indian towns in the Genesee country in 1779, and also when under the command of Brant and Sir John Johnson, the Indians subsequently avenged the invasion of Sullivan, by the fearful destruction they wrought in the valley of the Mohawk.

It was during this expedition that Cornplanter visited his father a second time. He was residing then in the vicinity of Fort Plain, and ascertaining where he lived, Cornplanter watched his opportunity and made his father a prisoner, but managed so adroitly, as to avoid recognition. He marched his sire ten or twelve miles up the river, and then stepped in front of him, faced about, and addressed him in the following manner:—

"My name is John O'Bail, commonly called Cornplanter. I am your son! You are my father! You are now my prisoner, and subject to the customs of Indian warfare. But you shall not be harmed: you need not fear. I am a warrior! Many are the scalps I have taken! Many the prisoners I have tortured to death! I am your son! I was anxious to see you, and greet you in friendship. I went to your cabin, and took you by force. But your life shall be spared. Indians love their friends and their kindred, and treat them with kindness. If now you choose to follow the fortunes of your yellow son, and to live with our people, I will cherish your old age with plenty of venison, and you shall live easy. But if it is your choice to return to your fields, and live with your white children, I will send a party of my trusty young men to conduct you back in safety. I respect you, my father: you have been friendly to Indians, and they are your friends."

The father preferred to return to his white children, and was therefore set at liberty, and escorted back in safety to his own home.

In another address to the governor of Pennsylvania, he used this language: "I will now tell you, that the Great Spirit has made known to me that I have been wicked; and the cause was the Revolutionary war in America. The cause of Indians having been led into sin, at that time, was that many of them, were in the practice of drinking and getting intoxicated. Great Britain requested us to join with them in the conflict against the Americans, and promised the Indians land and liquor. I myself was opposed to joining in the conflict, as I had nothing to do with the difficulty between the two parties.

"They told me they would inform me of the cause of the Revolution, which I requested them to do minutely. They then said it was on account of the heavy taxes, imposed on them by the British government, which had been for fifty years increasing upon them; that the Americans had grown weary thereof, and refused to pay, which affronted the king. There had likewise a difficulty taken place about some tea, which they wished me not to use, as it had been one of the causes that many people had lost their lives. And the British government now being affronted, the war commenced, and the cannons began to roar in our country.

"The white people who live at Warren, called on me, some time ago to pay taxes for my land; which I objected to, as I had never been called upon for that purpose before; and having refused to pay, the white people became irritated, called upon me frequently, and at length brought four guns with them, and seized our cattle. I still refused to pay, and was not willing to let the cattle go. After a time of dispute, they returned home, and I understood the militia was ordered out to enforce the collection of the tax. I went to Warren, and to avert the impending difficulty, was obliged to give my note for the tax, the amount of which was forty-three dollars and seventy-nine cents. It is my desire that the governor will exempt me from paying taxes for my land to the white people; and also cause that the money I am now obliged to pay, may be refunded to me, as I am very poor."

This appeal was brought before the Legislature of Pennsylvania, and an act was passed by which the chief was exonerated from the tax.

One writer speaks of him as possessing uncommon genius, a strong and discriminating mind, and as having the power of enduring great mental application. He anxiously inquired into the evidences in support of the scripture account of creation, and of the scheme of doctrines which Christianity unfolded.

President Alden of Alleghany college, speaks with delight of a visit he made to the old chief. He found him on the banks of the Alleghany, on a piece of first rate bottom land, a little within the limits of Pennsylvania. He was the owner of thirteen hundred acres of land, on a part of which stood his village, whose inhabitants gave signs of industry and thrift.

He found it pleasant to behold the agricultural habits of the place as appeared from the numerous enclosures of buckwheat, corn and oats. He also speaks of seeing a number of oxen, cows and horses; and many logs designed for the saw mill, and the Pittsburgh market. "Cornplanter had for some time been very much in favor of the christian religion, and hailed with joy such as professed it. When apprised of Mr. Alden's arrival he hastened to welcome him to his village, and to wait upon him. And notwithstanding his high station as a chief, having many men under his command, he chose rather, in the ancient patriarchal style, to serve his visitors himself; he therefore took care of their horses, and went into the field and cut and brought oats for them." [Footnote: Drake's book of the Indians.]

He died at his reservation March 7th, 1836, a hundred winters having passed over him, and was buried beneath the sheltering branches of a noble tree standing in his field. No other monument marks his grave.

CHAPTER XV

Change in Red Jacket's views—How caused—His opposition to Christianity— Visit of a Missionary—Missionary's speech—Red Jacket's reply—Unpleasant termination of the Council.

As time advanced, the mind of Red Jacket gradually receded from the favorable opinion he had entertained, with respect to the introduction among his people, of the customs of civilized life. Before this he regarded with favor the philanthropic designs of Washington and others, which contemplated such a change. But henceforth his influence and energies were uniformly exerted, in resisting any innovation, upon the anciently established usages of the Iroquois. Several causes seemed to influence such a result.

First of all was the condition of his people, as affected by the whites. They had been wasted and greatly enfeebled by the wars carried on between the whites, taking sides, as in the Revolution, against each other. And in their own conflicts, though in some instances successful, they had been so effectually overcome, that no hope now remained to them of resistance by war; no matter what combinations they might be able to effect among themselves.

A still deeper source of regret, was the loss of so large a portion of their wide and beautiful country. Since parting with it, swarms of settlers had been flocking to the more favored portions, and were irresistibly advancing to full and entire possession. The idea that they could have their country to hunt in, as well after it was sold as before, was rapidly dissipated by the busy sounds, all through the forest, of the woodman's axe, and by the roar of the stately trees, as they fell down before the enterprising pioneer. The Indian brooded over this in silence, while all of these sounds, delightful to the emigrant, were as a knell of death to his ear. The eloquence of Red Jacket had been exerted in vain, to arrest the progress of the white men. Onward they swept, bidding defiance to all the obstacles in their way. They were in possession of the ancient seats of the Iroquois. The red man's inheritance, was but a beggarly portion, when compared with his former princely domain. The thought of this weighed heavily upon Red Jacket's lofty spirit, and affected materially the disposition with which he regarded the white man.

He had observed also that the Indian had not been improved, but rather made worse by intercourse with the white man. He more readily acquired his vices, than his virtues.

The schools likewise that had been established among the Indians, had not been attended with very salutary results. And some of the Indian boys that had been sent to the schools of the whites, had failed to be qualified for usefulness among white men, and were unfitted in their tastes and habits for a life among the Indians. As was observed by Red Jacket: "they have returned to their kindred and color, neither white men nor Indians. The arts they have learned are incompatible with the chase, and ill adapted to our customs. They have been taught that which is useless to us. They have been made to feel artificial wants, which never entered the minds of their brothers. They have imbibed, in your great towns, the seeds of vices, which were unknown in the forest. They become discouraged and dissipated, —despised by the Indians, neglected by the whites, and without value to either,—less honest than the former, and perhaps more knavish than the latter." [Footnote: Washington had always been earnest in his desire to civilize the savages, but had little faith in the expedient which had been pursued, of sending their young men to our colleges; the true means he thought, was to introduce the arts and habits of husbandry among them.—Irving's Life of Washington.] Red Jacket was not alone in this opinion.

One of Cornplanter's sons, Henry O'Bail, had been educated in Philadelphia; but on returning to his people, became a drunkard, and was discarded by his father. He had other sons, but resolved that no more of them should be educated among the whites, for said he, "it entirely spoils Indian."

"What have we here?" exclaimed Red Jacket on one occasion addressing one of them. "What have we here? *You are neither a white man, or an Indian; for heaven's sake tell us, what are you?"*

But further than this, Red Jacket had witnessed among the whites so many evidences of deceit and fraud; he had so often seen the Indians circumvented by their avarice and craft, that he looked with suspicion even on their attempts to do the Indians good. The language of the Trojan patriot concerning the Greeks—represents very nearly the feelings he entertained toward the whites.

"Timeo Danaos et dona ferrentes.
"The Greeks I fear, e'en in the gifts they bear."

Hence Red Jacket began to look unfavorably on the attempts that had been made to civilize the Indian. He scorned to use the white man's axe, or hoe, or any implement of husbandry. He would not even use his language. Understanding well what was said to him in English, he spurned the idea of holding any communication with a white man, save through an interpreter. The Indian he looked upon as the rightful lord of this part of creation, the white man, as an intruder. The white man's ways were good for the white man; but in his view they would spoil the Indian. He believed that the peculiar characteristics of the Indian, were conferred on him by the Great Spirit for a wise purpose, and for his good, they needed to be maintained. Hence all the ancient habits of his people, he earnestly strove to preserve, and had it been in his power, he would have built a wall like the Chinese, to keep his people from meeting with, and being contaminated by the whites. He would frown contempt on the Indian, who used a stool or chair in his cabin, and no king in his palace, ever sat more proudly, or with greater dignity on his throne, than did Red Jacket on his bear-skin in his humble dwelling.

We can but admire in this, his independence of character; and when we reflect upon his conduct as influenced by the conviction, that such a course was essential for the good of his people, we may view it as meriting the praise of philanthrophy. Had he been as firm in resisting every enticement of the whites, he would have maintained a greater consistency, and himself attained a higher degree of excellence.

Red Jacket was equally opposed to the introduction of Christianity among his people. He looked upon the religion of the white man, with the same feeling of suspicion and distrust as everything else coming from that source. He had no evidence from experience, of the benefits that would arise to them from its introduction among them. On the contrary his convictions, arising from observation, were against it; because he saw his people were made worse, by associating with the whites. When asked on one occasion, why he was opposed to the coming of missionaries among his people, he replied,—"Because they do us no good. If they are not useful among the white people, why do they send them among the Indians?—If they are useful to the white people, why do they not keep them at home? They are surely bad enough, to need the labor of every one, who can make them better."

The Indians made no distinction between those who professed religion and those who made no profession. Their own religion was national. There was no division between the religious and irreligious. All were religious. In other words, they were all educated in the same faith, all united in observing the same religious rites, and all entertained the same religious belief, as had been handed down to them from their forefathers. This was salutary in promoting among them many virtues, worthy of commendation. They very properly estimated the value of religion, by the practical influence it exerted on those who received it. And they judged of the Christian religion, by the conduct and character of the nation that received and cherished it; who were nominally Christian.

Unfortunately for the success of Christianity among them, they had witnessed so much deceit and fraud, there were so many among the whites, who were ready to take advantage of them,—to make them drunk, and then cheat them, they were unable to perceive in what way the religion of the whites, from whom they had received such treatment, could be better or as good, even as their own. They had not learned to regard those only as Christians, who reduced the principles of Christianity to practice, and were not aware that as a system, it enforced only what was right, and tolerated no conduct that was wrong.

Hence in the efforts made to introduce Christianity among the Senecas, we find Red Jacket summoning the entire force of his influence, and eloquence in opposition to the measure.

The arrival among them of a missionary from Massachusetts, was the occasion of a forensic effort, which defines very clearly his position, and though it may have suffered, as did most of his speeches, from coming through an interpreter, it displays nevertheless, indications of deep thought, and of a high order of talent. It was regarded at the time as an effort of great ability, and is perhaps as fair a specimen of his oratory, as has come down to us from the past.

A council having been called to consider the matter, the missionary was introduced, who spoke as follows: [Footnote: The speech of the missionary is quoted from Col. Stone; the reply of Red Jacket from Drake, who is Col. Stone's authority for the same speech.]

"My Friends: I am thankful for the opportunity afforded us of meeting together at this time. I had a great desire to see you, and inquire into your state and welfare. For this purpose I have traveled a great distance, being sent by your old friends, the Boston Missionary Society. You will recollect they formerly sent missionaries among you, to instruct you in religion, and labor for your good. Although they have not heard from you for a long time, yet they have not forgotten their brothers, the Six Nations, and are still anxious to do you good.

"Brothers: I have not come to get your lands, or your money, but to enlighten your minds, and instruct you how to worship the Great Spirit, agreeably to his mind and will, and to preach to you the gospel of his Son, Jesus Christ. There is but one religion, and but one way to serve God, and if you do not embrace the right way, you can not be happy hereafter. You have never worshipped the Great Spirit, in a manner acceptable to him, but have all your lives, been in great errors and darkness. To endeavor to remove these errors, and open your eyes, so that you might see clearly, is my business with you.

"Brothers: I wish to talk with you as one friend talks with another; and if you have any objections to receive the religion which I preach, I wish you to state them; and I will endeavor to satisfy your minds, and remove the objections.

"Brothers: I want you to speak your minds freely; for I wish to reason with you on the subject, and if possible remove all doubts, if there be any on your minds. The subject is an important one, and it is of consequence, that you give it an early attention, while the offer is made you. Your friends, the Boston Missionary Society, will continue to send you good and faithful ministers, to instruct and strengthen you in religion, if on your part you are willing to receive them.

"Brothers: Since I have been in this part of the country, I have visited some of your small villages, and talked with your people. They appear willing to receive instruction, but as they look up to you, as their elder brothers in council, they want first to know your opinion on the subject. You have now heard what I have to propose at present. I hope you will take it into consideration, and give me an answer before we part."

The chiefs were in consultation for about two hours, when Red Jacket arose and spoke as follows:

"Friend and Brother: It was the will of the Great Spirit that we should meet together this day. He orders all things, and has given us a fine day for our council. He has taken his garment from before the sun, and caused it to shine with brightness upon us. Our eyes are opened that we see clearly; our ears are unstopped, that we have been able to hear distinctly the words you have spoken. For all these favors we thank the Great Spirit, and Him only.

"Brother: This council fire was kindled by you. It was at your request that we came together at this time. We have listened with attention to what you have said. You requested us to speak our minds freely. This gives us great joy: for now we consider that we stand upright before you, and can speak what we think. All have heard your voice, and all speak to you now as one man. Our minds are agreed.

"Brother: You say you want an answer to your talk before you leave this place. It is right you should have one, as you are a great distance from home, and we do not wish to detain you. But we will first look back a little, and tell you what our fathers have told us, and what we have heard from the white people.

"Brother: Listen to what we say.—There was a time when our fathers owned this *great island*. [Footnote: The term used by the Indians when speaking of this continent.] Their seats extended from the rising to the setting sun. The Great Spirit had made it for the Indians. He had created the buffalo, the deer, and other animals for food. He had made the bear, and the beaver. Their skins served us for clothing. He had scattered them over the country, and taught us how to take them. He had caused the earth to produce corn for bread. All this he had done for his red children because he loved them. If we had some disputes about our hunting ground, they were generally settled without the shedding of much blood. But an evil day came upon us. Your forefathers crossed the great water, and landed on this island. Their numbers were small. They found friends and not enemies. They told us they had fled from their own country for fear of wicked men, and had come here to enjoy their religion. They asked for a small seat. We took pity on them, granted their request, and they sat down amongst us. We gave them corn and meat; they gave us poison in return. [Footnote: Spirituous liquors.]

"The white people had now found our country. Tidings were carried back, and more came amongst us. Yet we did not fear them. They called us brothers. We believed them and gave them a larger seat. At length their numbers had greatly increased. They wanted more land;—they wanted our country. Our eyes were opened, and our minds became uneasy.

"Wars took place. Indians were hired to fight against Indians, and many of our people were destroyed. They also brought strong liquors among us;— they were strong and powerful, and have slain thousands.

"Brother: Our seats were once large, and yours were very small. You have now become a great people, and we have scarcely a place left to spread our blankets. You have got our country, but are not satisfied;—you want to force your religion upon us.

"Brother: Continue to listen. You say that you are sent to instruct us how to worship the Great Spirit agreeably to his mind, and if we do not take hold of this religion which you white people teach, we shall be unhappy hereafter. You say that you are right, that we are lost. How do we know this to be true? We understand that your religion is written in a book. If it was intended for us as well as you, why has not the Great Spirit given it to us, and not only to us, but why did he not give to our forefathers the knowledge of that book, with the means of understanding it rightly? We only know what you tell us about it. How shall we know when to believe, being so often deceived by the white people?

"BROTHER: You say there is but one way to worship and serve the Great Spirit. If there is but one religion why do you white people differ so much about it? Why are you not all agreed,—as you can all read the book?

"BROTHER: We do not understand these things. We are told that your religion was given to your forefathers, and has been handed down from father to son. We also have a religion which was given to our forefathers and has been handed down to us their children. We worship in that way. It teaches us to be thankful for all the favors we receive; to love each other, and to be united. We never quarrel about religion.

"BROTHER: The Great Spirit has made us all, but he has made a great difference between his white and red children. He has given us different complexions, and different customs. To you He has given the arts. To these He has not opened our eyes. We know these things to be true. Since He has made so great a difference between us in other things, why may we not conclude that He has given us a different religion according to our understanding? The Great Spirit does right. He knows what is best for is children; we are satisfied.

"BROTHER: We do not wish to destroy your religion, or take it from you. We only want to enjoy our own.

"BROTHER: You say you have not come to get our land or our money, but to enlighten our minds. I will now tell you that I have been at your meetings, and saw you collect money from the meeting. I cannot tell what this money was intended for, but suppose it was for your minister, and if we should conform to your way of thinking, perhaps you may want some from us.

"BROTHER: We are told that you have been preaching to the white people in this place. These people are our neighbors. We are acquainted with them. We will wait a little while, and see what effect your preaching has on them. If we find it does them good, makes them honest, and less disposed to cheat Indians, we will then consider again of what you have said.

"BROTHER: You have now heard our answer to your talk, and this is all we have to say at present. As we are going to part, we will come and take you by the hand, and hope the Great Spirit will protect you on your journey, and return you in safety to your friends."

Suiting the action to the word, he then drew near the missionary with the other chiefs, to take him by the hand; but he would not receive them, and rising hastily from his seat, said, there was no fellowship between the religion of God, and the works of the devil, and he could not therefore join hands with them.

When this was interpreted to the Indians, they smiled and retired from the interview, without saying anything further.

Such is the account given of this transaction. The termination is painful. It is a sad thought, that when the Indians had been invited to make known their objections, if they had any; and when they had been offered in good faith by this unlettered son of the forest, he should have been answered with so little patience or kindness. We hesitate not to say that the red man in this, manifested the better spirit.

Mr. Crane afterward regretted the course he had taken, saying, "he supposed by shaking hands with them, they would regard it as signifying his approval of what they had said."

CHAPTER XVI

Tecumseh and Indian confederation—Aid given by Elskawata—Doings at the Prophet's Town—Great Indian Council at the West—Red Jacket's claim for precedence of the Senecas—His adherence to the United States—Hostilities encouraged by British Agents—Warriors gathered at Prophet's Town—Visited by Gen. Harrison at the head of his troops—Hostilities disclaimed— Surprised by a sudden attack—Indians defeated—War proclaimed against England—Indians take sides—Unfavorable commencement—Different successes —Part taken by Red Jacket.

Sixteen years had intervened since the treaty of peace, concluded with the Indians at Greenville, by General Wayne in 1795. During this time friendly relations had been maintained with the various Indian tribes, who were in correspondence with the United States. This period had not closed, however, ere the ambitious designs of an active and influential chief, began to wear the appearance of open hostility.

The possession of rare mental endowments, together with physical qualifications, that were the means of extending his renown, as an intrepid brave, far beyond the boundaries of his own tribe, rendered the name of Tecumseh, a rallying word for the dusky warriors, even among the remote wilds visited by the Indian. Tecumseh entertained the ambitious project, at various times a favorite design with the Indian, of uniting all their tribes at the West and South-west, in one strenuous endeavor, to resist the further advance of the whites into their country, and of forming here a confederacy, similar to that which had existed among the Iroquois.

In these views he was greatly assisted and strengthened by the influence and efforts of his brother, Elskawata.

Elskawata, on the death of Penagashega, an aged and revered prophet, very adroitly assumed the sacred office of this Indian saint, and began to proclaim himself, as a delegated messenger of the Great Spirit to his people.

He commenced his career among the Shawnees, the people of his tribe, as early as 1805. But not content with so narrow a sphere for his endeavors, he went from tribe to tribe, and assembled as he was able, different nations, that he might make known to them the important instructions, he had been divinely authorized to communicate.

For a long time his efforts wore the appearance of a religious, and pacific character. He proclaimed the high superiority of the Indians over the whites, and of his own tribe among the Indian tribes. He declared it to be the will of the Great Spirit, that the Indians should abandon the use of intoxicating drinks, refrain from intermarrying with the whites, live at peace with each other, have their property in common, and maintain their customs, as they had been anciently established. At a later period he affirmed with much solemnity, that he had received power from the Great Spirit, to cure all diseases, confound his enemies, and stay the arm of death, in sickness, or on the field of battle.

As time advanced, the prophet passed from nation, to nation, artfully sustaining his assumptions, and proclaiming his doctrines. He gathered around him adherents from various tribes, encouraged pilgrimages to his camp, became conspicuous in all their general councils, and extended his influence to the various Indian towns, in the vicinity of the northern lakes, and on the broad plains, watered by the Mississippi and its branches. He could now, as he did, forward very effectively the ambitious views of his brother Tecumseh.

From the Prophet's town, which was established on the banks of the Wabash, near the mouth of its tributary the Tippecanoe, as early as 1808, a correspondence was kept up with the numerous tribes at the North and West, and means were taken also to extend the combination they were forming, to the Cherokees and other nations of the South. Runners were sent as far even as the country of the Senecas, and the Iroquois in New York and Canada, were solicited to join the Great Western Confederacy.

Connected with this movement was the holding of Indian councils, at different places in the West. A very large council, was held at or near Detroit, which embraced in it deputations from the most distant tribes. A strong deputation was sent from the Senecas, with Red Jacket at its head.

At the opening of this council a question arose as to precedence in debate, which is said to have been the occasion of one of Red Jacket's most effective and brilliant speeches, and was the means of securing for himself and fellow delegates, the high position he ever claimed, as belonging rightfully to his nation.

The right of precedence was claimed by the Wyandots, a large and powerful nation, which for a long time, had been pre-eminent among the Western tribes. To them had been committed for preservation and safe keeping, the Great Belt, the symbol of a previous union among the tribes. It had been used in gathering them to form their league, to resist the settlements of the whites north and west of the Ohio river. The concert of action among the Indians, in the wars at the West between 1790 and 1795, is to be traced to this league. To the Wyandots also had been given the original duplicate of the treaty of peace, concluded at Greenville in 1795.

Hence the claim they presented to precedence at this council; a claim which was eloquently supported by their most able chiefs.

This claim was denied by Red Jacket, who maintained that the place in question belonged rightfully to the Senecas, and sustained his position by a reference to facts and usages in the past, which displayed a minute and accurate knowledge of the history of the different Indian tribes, that surprised as well as delighted his hearers. His speech was characterized throughout by great ability, and displayed such a power of oratory, particularly of invective, as to excite the wonder of all present, who could understand his language, and comprehend the force of his allusions. His effort was entirely successful. No attempt was made at reply. The first rank after this, without further hesitation, was given to the Senecas.

It is due to the memory of Red Jacket, who has been, called *double tongued and deceitful*, to state that from the time he fully gave his adherence, he never swerved from his allegiance to the United States. Ever afterward he was their faithful friend and ally. The impatient affirmation of Brant, that "Red Jacket had vowed fidelity to the United States, and sealed his promise, by kissing the likeness of General Washington," though in a measure true, as expressive of his fidelity, had never any occasion to be qualified, by a statement to the contrary.

During the present council, his views were in opposition to those generally entertained and expressed, and no consideration availed with him, to break faith with the United States. He had before this notified the Indian agent of the formation of another league, and of the avowedly warlike purpose of certain Indian councils, that had been held at the West.

Early in the year 1810, at the head of a delegation of his people and accompanied by the agent, and Captain Parrish as interpreter, he visited the city of Washington, and informed our government of the hostilities that were in contemplation, and of the efforts of his people to secure peace.

The pacific councils of Red Jacket were of little avail. The warlike agitation was continued. The retreat of the Prophet on the banks of the Wabash, became not less noted for warlike exercises, than for its religious harangues. The minds of the Indians were already ripe for an outbreak, whenever a sufficient pretext should offer. The visit of Tecumseh at Vincennes in the summer of 1810, with three hundred well armed warriors, and his haughty and insulting bearing toward Governor Harrison, indicated clearly, the hostile spirit that was rife among them.

Not long after this, the report came that a thousand warriors awaited his command, in and about the Prophet's town. So large a horde of Indians together, without the means of support, and practicing themselves in the arts of war, were viewed with suspicion. Charity must have been blind, to have supposed they were assembled merely for the purpose of devotion. Frequent plundering, midnight arson, and occasional massacres in frontier settlements, proclaimed the fact, that hostilities had already commenced, and that our people in this region needed protection.

The Indians were greatly encouraged in their warlike feeling, by the intercourse they constantly maintained with the British Indian Department. The British Fur Company also by her traders, had correspondence with the leading men of all these Western and North-western tribes, and this intercourse resulted in holding the Indians more firmly, in alliance with the English. The desire they entertained for dominion on this continent, led them to encourage the Indians, in their effort to hold in check the settlements of the United States, that were pushing their way westward. Thus countenanced and encouraged, the Indians became more determined and bold in their hostility.

These threatening indications, coming to the knowledge of our government, General Harrison was directed to go with an armed force to the Prophet's town, and his visit resulted in the battle of Tippecanoe, fought on the seventh of November, 1811.

His officers desired him to attack the town on the day before, but wishing to avoid fighting if possible, and having been met by several chiefs, who disclaimed having hostile intentions, and offered submission and peace, he made a careful survey of the country, and selecting an advantageous position, encamped for the night.

At an early hour in the morning they were furiously assailed by the Indians, who had stealthily crept up very near without being observed. A bloody and, for a time doubtful, engagement ensued, but at length the Indians were repulsed and a decisive victory gained.

The Prophet was securely stationed on an adjoining eminence during the battle, and the American bullets having a more powerful effect upon the Indians than they had been led to anticipate, a runner was sent to him with the intelligence. He was engaged singing very piously, one of his old war songs. When told what was taking place, he said, "Go,—fight on: it will soon be as I have said;" and commenced singing again more loudly. [Footnote: The Prophet had assured them that the Americans would not be successful. That their bullets would not hurt the Indians, who would have light while their enemies would be in darkness.—Life and Times of Wm. H. Harrison.]

Tecumseh was absent when the battle was fought, being engaged in forwarding his designs among the Indian tribes at the South. He was disappointed and grieved with the result, regarding the battle as premature, and tending very much to thwart the purposes he had in view.

He awaited a more favorable turn in the wheel of fortune, and thought this would come with the war anticipated between England and the United States. Difficulties, growing out of the right assumed by the former, of boarding American vessels, to discover and remove any English sailors belonging to the crew, which frequently resulted in seizing American seamen and forcing them into the British navy, had now assumed so formidable an aspect, as to call forth from our government a proclamation of war against England, issued on the 19th of June, 1812.

In anticipation of this event, as well as after it, means were employed by the agents of Britain, to secure the services of the Indians during this contest. The opportunity was gladly welcomed by the Miamis, Shawnees and other Indian tribes, who had recently been severely chastised by General Harrison. The Mohawks and other Indians in Canada were also induced to take up the hatchet, and efforts were made to influence such of the Six Nations, as resided within the state of New York, to take sides with the British in this war, but they were not successful.

The United States, instead of seeking among the Indians recruits for their army, advised the Senecas, and other tribes of the Iroquois within their borders, to remain neutral. A council was convened by the Indian agent, Mr. Erastus Granger, for the purpose of spreading the whole matter before them. It resulted in securing from them a pledge of neutrality. So well convinced were they of the wisdom of this course, they determined to send a deputation of their brethren to Canada, to dissuade them if possible, from taking any part in the war. It was sent, but did not accomplish the end desired; the Mohawks had fully resolved upon engaging in the contest.

It was difficult however, for the Senecas to enforce their decision upon their young braves, who were made restless by the sound of war, and were eager to engage in it; yet their sympathies were with the United States. The stirring music, martial array, noise and pomp of war, wrought so effectually on their minds, they would fain have persuaded their nation to declare war on their own account. The circulation among them of a rumor that the British had taken possession of Grand Island, a part of their own domain, led them to convene a council, which Mr. Granger was invited to attend, and after stating the case to him, Red Jacket declared the purpose of the Senecas in the following language:—

"BROTHER: You have told us, that we have nothing to do with the war, that has taken place between you and the British. But we find that the war has come to our doors. Our property is taken possession of, by the British and their Indian friends. It is necessary for us now to take up the business, defend our property, and drive the enemy from it. If we sit still upon our seats, and take no means of redress, the British according to the customs of you white people, will hold it by conquest. And should you conquer the Canadas, you will claim it on the same principle, as though you had conquered it from the British. We therefore request permission to go with our warriors, and drive off those bad people, and take possession of our lands."

Their request was granted, and the chiefs regarding themselves as an independent nation, issued a formal declaration of war, against the provinces of Upper and Lower Canada, and summoned their warriors to protect their rights and liberties, with the Americans.

Four hundred warriors, armed and painted, and ready for the field, answered to this call, led by the brave though now aged Farmer's Brother, who was said by Colonel Worth, to have been "the noblest Indian in form and mould, in carriage and in soul, of that generation of his race." [Footnote: Col. Worth as given by Col. Stone.]

The principal scene of war at this time was on our north-western frontier. Its commencement had been disastrous. The capture of Mackinaw, Chicago, and Detroit, attended by the surrender of General Hull, commander of the American forces at the latter place, spread a feeling of insecurity and dismay all along our western frontier settlements. For an immense extent they were without protection. But new troops were raised and brought on to the field, under the wise conduct of General Harrison, and the signal naval victory of Commodore Perry on Lake Erie, September 10, 1813, and the equally decisive battle on the river Thames, in the October following, very materially improved the prospect of the American arms. After this battle, the Western Indians were disposed to entertain propositions of peace.

Thus far they had proved to be efficient allies of the British, increasing their force by an addition of nearly eighteen hundred, commanded by the renowned Tecumseh, who had been called the "Indian Bonaparte." His pre-eminence among them was now widely acknowledged, and he swayed by his influence a greater number of warriors, than any Indian of his time. Before engaging in the Battle of the Thames, he seemed to have a presentiment of his death. He said to the chiefs about him, "brother warriors, we are about to enter into an engagement from which I shall never come out,—my body will remain on the field of battle." [Footnote: Life of Tecumseh, by Benjamin Drake.]

His prediction was verified; as marking the field of strife where the Americans were victorious, the ashes of this celebrated warrior here repose, near the borders of a willow marsh, the willow and the wild rose weaving a chaplet over his grave.

The Indians who had volunteered their services in the American army, were first employed in the gallant defense made at Black Rock, during the month of July, of this same year, 1813.

A surprise party from the British head-quarters at Lundy's Lane, was sent against the American stores, collected at Black Rock and at Buffalo, and were not at this time strongly guarded. They were successful in their first attempt, but were in turn unexpectedly met by the adroit management of General Porter, under whose supervision the forces in this vicinity had been placed, who rallied volunteers at Buffalo, turned back the retreating garrison, and by a well planned attack, succeeded in driving the enemy from the post they had taken a short time before.

The Indians were soon after in another engagement, in the vicinity of Fort George, and from an official report made at the time, it appears that this formed a part of Red Jacket's military experience. A company of volunteers and Indians commanded by Major Chapin, to which was added a force of about two hundred regulars under Major Cummings, amounting in all to about five hundred, the whole being under the direction of General Porter, proceeded to attack the British and Indian encampment, and were entirely successful.

In an official statement of this affair given by General Boyd, then commanding the post at St. George, he says:—"Those who participated in this contest, particularly the Indians, conducted with great bravery and activity. General Porter volunteered in the affair, and Major Chapin evinced his accustomed zeal and courage. The principal chiefs who led the warriors this day were Farmer's Brother, Red Jacket, Little Billy, Pollard, Black Smoke, Johnson, Silver Heels, Captain Half Town, Major Henry O'Bail, and Captain Cold, who was wounded. In a council held with them yesterday, they covenanted not to scalp or murder; and I am happy to say, that they treated the prisoners with humanity, and committed no wanton cruelties on the dead."

CHAPTER XVII

Taking of Fort Erie—Battle of Chippewa—Service rendered by the Indians— General Porter's account of the campaign—Red Jacket commended—Withdrawal of Indian forces—Other successes—Conclusion of peace.

Directly across from Buffalo, at the head of Niagara river, on the Canada side, stood Fort Erie. Chippewa, at that time head-quarters of the British army, was eighteen miles below, on the same side. Fort Erie was garrisoned by about one hundred and seventy men; at Chippewa and within available distances from it, was stationed the principal part of the British force in this region.

The plan with which it was proposed to open the campaign of 1814, contemplated an attack on both of these places.

The campaign of the previous year, though favorable in a good degree, did not close with entire success to the American arms. The idea was entertained of descending the St. Lawrence, with a view of capturing Montreal, a design which signally failed. Taking advantage of the feeble defense of our frontier, by the withdrawal of the regular troops for the purpose named, the enemy, on the 18th of December, surprised and took Fort Niagara, and sweeping along our frontier settlements on the Niagara river, ravaged the country by fire and sword, as they passed rapidly on, carrying the works at Lewiston, and Manchester, and laying in ashes the thriving villages of Black Rock and Buffalo. They burned also without opposition, a village of the Tuscaroras.

The voice of Red Jacket was thereupon heard, arousing his people again, to the necessity of taking up arms. And as the result, about six hundred warriors, mostly from the Seneca nation, were in readiness to offer their services, at the opening of the present campaign.

Buffalo was the appointed place of rendezvous, and on the first of July, General Brown, who commanded our forces, regarding them as sufficient to warrant the commencement of the plan of operations, began by reconnoitering Fort Erie. During the night of the second of July, General Ripley, with a part of his brigade, embarked in boats, with a view of landing on the opposite shore, one mile above the Fort, at about day break the next morning.

General Scott with his brigade was to cross the Niagara river, through a difficult pass in the Black Rock Rapids, and make a simultaneous landing below the fort. The two brigades enclosing the fort, could prevent the escape of the garrison, until artillery to reduce it, should be brought from Buffalo.

General Scott with his usual promptness, made good his landing, and was on the ground at the hour appointed, and by the aid of a few Indians and volunteers who accompanied him as guides, invested the fort, so as to secure its garrison. General Ripley though prompt in his departure, was delayed in reaching his position, by a dense fog which misled his pilots.

As the sun rose the British commandant and his officers, could see the busy operations going on in ferrying across from Buffalo, artillery, Indians and soldiers, with their various preparations of war. They discovered also how completely they were invested. At the demand of General Brown, without firing a gun or making any attempt at resistance, the fort and garrison were surrendered.

This part of their enterprise being accomplished, they next turned their thoughts toward Chippewa.

The Chippewa or Welland river, is a considerable stream not far from one hundred yards wide, and from twelve to twenty feet deep, entering the Niagara at right angles, as it flows in from the west.

On the north or left bank of this stream, near its mouth, the British army had its station and defenses, consisting of two block houses, connected and flanked by a parapet.

Street's creek was two miles above, or south of this, a small sluggish stream, which enters the Niagara in a direction parallel with the Chippewa. The mouth of this creek was selected by the American commanders, as affording a favorable position for their army before the battle.

On the evening of the same day of the capture of Fort Erie, General Scott with his brigade and Towson's artillery, proceeded down the river on his way toward Chippewa, and on the morning of the fourth, encamped in the open field, on the south side of Street's creek, having driven in some advanced posts of the enemy. In the evening he was joined by General Brown, with General Ripley's brigade, which took post in the same field, in rear of General Scott.

General Porter with the Indians, and Pennsylvania volunteers, crossed the Niagara at Black Rock during the night of the fourth, and on the morning of the fifth, marched for the camp, arriving there at about noon.

The two armies nearly equal in numbers, and well qualified by their thorough equipments, and the skill of their commanders, to harm each other effectually, were now encamped with only two miles, and the two streams, on whose banks they rested, between them.

But though thus near, intervening objects prevented their seeing each other. Between them was a strip of woodland about one-fourth of a mile in width, extending from the forest on the west, near to the bank of the river, where it was cleared for the public highway. This effectually shut out from the view of the other, the manoeuvres of each army.

The Indians and militia from the British army infested these woods, and became annoying to our forces. General Porter being well acquainted with the country, and having charge of the Indians, was requested to take them, and a part of his Pennsylvania volunteers, and dislodge this portion of the enemy; General Brown assuring him, that none of the British regulars would be found south of the Chippewa on that day, and promising him in case of so improbable a contingency, the support of General Scott's brigade.

At about three o'clock of the same day of his arrival, General Porter formed his men, half a mile in rear of the main camp, into single or Indian file, placing the Indians on the left, and a part of the Pennsylvanians on the right.

"Thence he marched into the woods in the same order, in a line at right angles to the river, until the whole Indian force was immerged in the forest, leaving the white troops in the open field; they had only to halt and face to the right, when the whole were formed in line of battle, three-fourths of a mile long and one man deep, looking in the direction of Chippewa. Red Jacket was placed on the extreme left of the line, and General Porter took his station on the margin of the woods between his white and red troops, accompanied by Captain Pollard, a Seneca chief, who, in this campaign, was considered first in command among the Indians; Colonel Flemming, the Quarter-master of the Indian corps, Lieutenant Donald Fraser his aid, and Henry Johnson his interpreter. He was also accompanied by Major Jones, and Major Wood of the Engineers, as volunteers; and was supported by a company of regular infantry, marching in column in rear, as a reserve.

"The Indians were commanded by their war-chiefs, who were indulged in their own mode of conducting the attack, marching about twenty yards in advance of the warriors of their respective tribes. General Porter having sent out scouts to reconnoiter the enemy, the march was commenced by signal, and proceeded at first with great stillness and caution. The chiefs have signals, by which, on the discovery of any circumstance requiring consultation, or change of route or action, they convey notice through their ranks with great celerity, on which the whole line of warriors drop instantly on the ground, and remain there until further orders. Two manoeuvres of this kind occurred on the march, the first of little moment, but the second communicating through the scouts, the exact position of the enemy, who, apprised of their assailants' approach, lay concealed in a thicket of bushes, along the margin of Street's creek.

"A consultation was thereupon held, and new orders given, the purport of which was to change the line of march, so as to meet the enemy to more advantage, to increase the speed as much as was consistent with the preservation of order, and to receive their first fire, but not to return it except singly, and when it could be done with certain effect, and then to raise the war-whoop, pursue, capture, and slay as many as practicable, until they should reach the open ground in front of Chippewa, and thence return to camp.

"The march was accordingly resumed, the fire of the enemy received, and a rush accompanied with savage yells made upon them, and continued for more than a mile, through scenes of frightful havoc and slaughter, few only of the fugitives offering to surrender as prisoners, while others, believing that no quarter would be given, suffered themselves to be cut down with the tomahawk, or turning back upon their pursuers, fought hand to hand to the last.

"On reaching the open field in front of Chippewa, the assailants were met by a tremendous discharge of musketry, by which the warriors, who were principally in front, were thrown back upon the volunteers and reserve, who for want of equal speed were a short distance in the rear. Presuming that the fire had come from the enemy he had been pursuing, and who had rallied on reaching the open ground, General Porter made an effort, not without success, to reform his line with volunteers, reserve and a portion of the warriors; but on again advancing to the margin of the woods, found himself within a few yards of the whole British regular army, formed in line of battle, and presenting within a given space at least three men fresh from their camp, to a single one in his own attenuated and exhausted line. After receiving and returning two or three fires, the enemy rushed forward with charged bayonets, when hearing nothing from General Scott, he gave the order to retreat and form again on the left of General Scott's brigade, wherever it should be found.

"It appears that the British commander had resolved on making a general attack, that day, on the American camp; and in execution of this purpose had marched his whole force across the Chippewa, a short time before General Porter entered the woods with the Indians; and having sent forward his Indians and militia, which was the British force met in the woods, to commence his attack on the left flank of the Americans, formed in the meantime his battalions of regulars on the plain, under cover of the strip of woodland which divided the two camps, with his artillery on his left, near the gorge occupied by the road along the bank of the river; ready to act the moment the effect of the flank attack should be developed.

"The repulse of General Porter's command was thus effected by the main body of the British army, while General Scott's brigade was more than a mile in the rear, and had not yet crossed the bridge over Street's creek.... In a retreat of a mile in a diagonal direction to the right, so as to uncover the enemy to the fire of the American line, then just beginning to form, they gained but little distance on the British columns, who were in hot pursuit. When General Porter and his staff arrived at Street's creek, they were met by Major Jessup's battalion, then in the act of taking its position, which was on the left, and a short distance from the remainder of General Scott's brigade; and the volunteers fatigued as they were, aided Major Jessup's evolutions, which were executed with great order and celerity, by breaking down the fences to enable him to pass from the road bordering on Street's creek, to his position in the field.

"Nothing could exceed the coolness and order with which General Scott's brigade crossed the bridge and formed its line, under the galling fire of the enemy's artillery, and the headlong approach of his infantry, who, when only fifty yards distant, were received by a tremendous discharge of musketry from the American line, which forced them to fall back for a considerable distance. But they speedily rallied and advanced again, when they were met in the same gallant manner; and they thereupon fled, with as much precipitation as they had entered it, not halting until they had recrossed the Chippewa and destroyed their bridge.

"General Scott pursued them around the point of woods, beyond which he could only advance in face of their batteries, and these he could not reach by reason of the intervening river. He therefore deployed to the left, and forming a line in the open field, in front of Chippewa, directed his men to lie down with their heads toward the batteries, the better to avoid the effect of their fire.

"The battle between the regular troops, was but of a few minutes duration, with the exception of the artillery, which on both sides was earliest and longest engaged, and served with the most destructive effect; Colonel Towson occupying the right of the American line, on Street's creek, and the British artillery the left of theirs, at the point of woods, and both commencing with the first movements of the regular troops.

"Immediately after the two lines had encountered on Street's creek, a magnificent charger completely caparisoned, but without a rider, was seen prancing and curveting in the centre of the battle field, and endeavoring to make his escape through the American line to the rear. Presuming that he belonged to some officer who had fallen, he was forthwith secured by the servant of General Porter, and immediately mounted by the General, to whom he was a most acceptable acquisition, after the labors of the day, which he had performed on foot.

"Riding up to General Brown, who was also in the midst of the action, General Porter received his orders to march with the two hundred Pennsylvanians, who had been left in camp, to the support of General Scott; which orders were promptly executed by following General Scott's brigade around the point of woods, receiving the fire of the British batteries, and taking post on his left, with the men in the same recumbent position. Here they awaited the arrival of General Ripley's brigade, which on the first discovery that the whole British army was in the field, had been ordered to make a detour through the woods, and attack the enemy's right. They soon came up, in the same muddy plight with the volunteers and Indians, who had previously traversed the same ground; when the whole army at about sundown quietly retired to their camp, on the south side of Street's creek.

"And thus ended the battle of Chippewa, which probably produced more important results in favor of the American arms, than any other engagement by land in the course of that war; although there were several battles fought on the Niagara, if not elsewhere, during the same campaign, exhibiting a greater number of combatants engaged, a larger number of slain, and a result equally creditable to the gallantry and good conduct of the American soldiers.

"The first advantage gained was in driving from the British army those troublesome enemies, their Indian allies, who had been the terror of our troops in the west, during all the preceding stages of the war, and had kept the camps of General Dearborn, General Lewis, and General Boyd, in a perpetual panic during the campaign of 1813. Terrified and disheartened by the reception they met with at Chippewa, they fled from the battle field to the head of Lake Ontario, a distance of thirty miles, without halting, and never again during the remainder of the war appeared in the British camp." [Footnote: Colonel Stone's Life and Times of Red Jacket. Mr. Stone refers to General Porter, as his authority, representing him as having voluntarily prepared the account given of this campaign.]

The Indians during this engagement performed a most important service. Their conduct was highly commended by General Porter. Speaking of those under his command, General Porter says: "The great body of warriors as well as volunteers, engaged in the opening attack, fought with boldness, not to say desperation, unsurpassed by any other troops, until they were placed in a situation where it would have been madness not to retreat."

The part Red Jacket took in this battle, though by no means conspicuous, was such as to call forth from an early biographer the affirmation, that "he displayed the most undaunted intrepidity, and completely redeemed his character from the suspicion of that unmanly weakness, with which he had been charged in early life; while in no instance did he exhibit the ferocity of the savage, or disgrace himself, by any act of outrage towards a prisoner, or a fallen enemy."

The same writer adds: "His therefore was that true moral courage, which results from self respect, and the sense of duty, and which is more noble, and a more active principle, than that mere animal instinct which renders many men insensible to danger. Opposed to war, not ambitious of martial fame, and unskilled in military affairs, he went to battle from principle, and met its perils with the spirit of a veteran warrior, while he shrunk from its cruelties with the sensibility of a man and of a philosopher." [Footnote: Life of Red Jacket. McKenny's Indian Biography.]

Red Jacket as a civil officer was not called to take so prominent a place on the field of battle, as the war chiefs. Yet in all of their deliberations, which were frequent during the campaign, he could act as their counsellor, as he did on every such occasion. He was uniformly their principal orator, and his manner on these occasions is represented as being "graceful and imposing in the eye of every beholder, and his voice music, especially in the ears of his own people. He had the power of wielding them at will, and the soul stirring trumpet could not produce a more kindling effect in the bosoms of a disciplined army, than would his appeals upon the warriors of his race." [Footnote: Col. Stone's Life of Red Jacket.]

That the battle of Chippewa was particularly severe to the Indian forces engaged in it, may be inferred from the fact that the British Indians retreated not only beyond the Chippewa, but stayed not until they had gone thirty miles further. The battle ground was strewed with many of their number who had been slain. Two, who had been mortally wounded, and were still alive, were despatched by a party of New York Indians, who were looking for the bodies of their fallen friends. Being reproached for their conduct in taking the life of an unresisting foe, one of them replied, in a manner that indicated evident sorrow for the deed done, "That it did seem hard to take the lives of these men, but they should remember that these were very hard times." [Footnote: Col. Stone.]

The sight of slain warriors was far from being a pleasing object for Red Jacket to behold, and having ever been opposed to his people engaging in contests that did not really concern them, he proposed now that the Indians had helped chastise the British for burning one of their villages, and as they were no longer on Indian ground, that they should withdraw from a further participation in the war, in case they could prevail on their Canadian brethren to do the same.

With the consent and approval of General Brown, a deputation of two brave and influential chiefs was sent to the Indians, who had fought with the British, with this in view. They were successful in persuading them to enter into this arrangement. The Indians therefore after this retired to their villages, with the exception of a few young braves, with whom the love of war, was a more potent influence, than the counsels of the aged and more considerate of their nation.

Soon after the battle, our army forced a passage across the Chippewa, and after a short engagement the enemy gave way, and retired to Lake Ontario. Our army continued its march down the Niagara river, destroying some of the British works on their way.

With new forces brought into the field, General Drummond took command of the British, and on the 25th of July the two armies met again, and there was a hard fought, but not very decisive battle, at Lundy's Lane, near Niagara Falls. The American army soon after fell back to Fort Erie. A British force of five thousand advanced and laid siege to the Fort, making a vigorous assault on the 15th of August. They were repulsed with a loss of a thousand men. Later, General Brown issued from the fort and gave them so stunning a blow as caused them to relinquish the siege.

Other successful engagements during the year, ending with the signal victory at New Orleans under General Jackson, inspired greatly the hopes of the American people, and served likewise to repress the ardor of their opponents; which led to the return of peace with England, which was concluded at Ghent on the 24th of December, 1814.

CHAPTER XVIII

Pre-emptive right to the Indian Reservations sold to Ogden and Company— Council to obtain the lands—Mr. Ogden's speech—Red Jacket's reply— Indians refuse to sell—Another Council called—Account of it by Hon. Albert Tracy—Various utterances of the orator on that occasion—Indians' appeal to the governments of the United States and New York—Noble response of Governor De Witt Clinton of New York—Final success of the Ogden Company.

Though the Indian lands within the state of New York, had now been narrowed down to a comparatively small compass, there were not wanting those who would take from them, the remaining portion of their ancient inheritance. The preemptive right to their reservations was sold by the Holland Land Company, to Colonel Aaron Ogden and others, who were known as the Ogden Company. The efforts of these gentlemen to induce the Indians to dispose of their reservations, resulted in calling several Indian councils, at which Red Jacket was the prominent speaker, and in which the entire force of his great powers was summoned, to withstand and thwart their endeavors.

A council for this purpose was convened at their village near Buffalo, during the summer of 1819. The Hon. Morris S. Miller of Oneida, was present as a commissioner on the part of the United States; and the Hon. Nathaniel Gorham of Canandaigua, represented the interests of the state of Massachusetts. Captain Parrish of Canandaigua, and Captain Horatio Jones of Genesee, were present as interpreters.

As it was known Red Jacket was to speak in opposition to the interests of the Land Company, the occasion drew together a large concourse of people; pale faces as well as red, who were interested in the result of the negotiations contemplated, as also by a desire to hear the speech of the distinguished orator of the Senecas. Of this Colonel Stone remarks: "No subsequent assemblage of the Indians within the state of New York, has presented so numerous and imposing an array, nor is it likely that so many of them will ever again meet, on the soil of their fathers."

A gentleman who was present at this treaty by the invitation of a friend, speaks of it, in the following terms:—"My friend and myself having arrived on the ground at an early hour; we saw at a little distance from the wigwams, a group of Indians, under the shade of a cluster of plum trees, lying on the ground. Among these were a number of chiefs, of whom in a conspicuous place, was Red Jacket, apparently in deep thought, with a pile of little small sticks, two or three inches long, before him.

"I inquired of a gentleman who was conversant with Indian proceedings, what Red Jacket was doing? He replied that he was studying his speech, and advised us to retire, as he perceived it disturbed him. About this time the commissioners, Governor Ogden, his friends, and the two celebrated Indian interpreters, Parrish and Jones, and a large concourse of people, gentlemen and ladies, began to assemble under another cluster of trees, where benches had been prepared in two parallel lines, with a wide space between, and seats across the upper end, for the commissioners. The long seats were occupied on the right by Messrs. Ogden, their officers, and other gentlemen and ladies; the left by Red Jacket, a large number of chiefs, and other Indians.

"There was order, dignity, and perfect silence. The contest soon commenced. Governor Ogden, a dignified, fine looking man, rose and opened the case. Mr. Parrish, a man of large stature, stood up at the same time, and interpreted it to the Indians, sentence by sentence." [Footnote: Author's Scrap Book.]

The object was to buy the Indian title, as they had already brought the pre-emptive right. Governor Ogden told them it was the wish of their great Father, the President of the United States, that they should sell these lands, and go down to a reservation on the Allegany river, where they could live in peace, and have a good foothold forever; and used various arguments in favor of such a course.

After Governor Ogden had finished his speech, Red Jacket rose with a great deal of composure, and adjusting his belt of handsome wampum, and looking to the sky for a moment spoke. Mr. Parrish interpreted: "Red Jacket says he thanks the Great Spirit that we are all alive and here this pleasant day." He then addressed the commissioners, answering all the statements and arguments of Governor Ogden in their order, unfolding a long roll of parchments attached together, of treaties that had been made at different times by the United States, with the Six Nations. They had been preserved in good order. He pointed to the dates, and to the substance of the treaties from time to time, with great accuracy, as appeared from the interpretation, answering Governor Ogden with the most forcible arguments, interspersed with wit and humor. His speech on this occasion, as quoted by Col. Stone, is as follows:

"Brother: We understand that you have been appointed by our great Father, the President, to make these communications to us. We thank the Great Spirit for this pleasant day given us for our reply, and we beg you to listen.

"BROTHER: Previous to your arrival at this council fire, we were told that our great Father had appointed a commissioner to meet us. You have produced your commission, and it has been read and explained to us. You have also explained the object of your mission, and the wishes of the President in sending you to the council fire of the Six Nations. We do not doubt that the sealed document you produced, contained the words of the President, our great Father. When first informed of your appointment, we supposed that you were coming to meet us on a very different subject. Since the war of the Revolution, we have held various councils with our white brothers, and in this same manner. We have made various speeches, and entered into several treaties, and these things are well known to our great Father; they are lodged with him. We, too, perfectly understand them all. The same interpreters were then present as now. In consequence of what took place during the late war, we made it known to our great Father, through our interpreter, that we wished to have a talk. Our application was not complied with. We sent a messenger to brighten the chain of friendship with our great Father, but he would not meet around the council fire, and we were disappointed. We had supposed that the commissioner he has now sent, came forward to brighten the chain of friendship, to renew former engagements. When we made a treaty at Canandaigua with Colonel Pickering in 1794, we were told, and thought that it was to be permanent, and to be lasting, between us and the United States forever. After several treaties had been entered into under our great Father, General Washington, large delegations from the Six Nations were invited to meet him. We went and met him in Philadelphia. We kindled a council fire. A treaty was then made, and General Washington then declared that it should be permanent between the red and white brothers; that it should be spread out on the largest and strongest rocks, that nothing could undermine or break; that it should be exposed to the view of all.

"BROTHER: We shall now see what has been done by the United States. After this treaty had been formed I then said that I did not doubt, but that the United States would faithfully perform their engagements. But I told our white brothers at that time, that I feared eventually they would wish to disturb those contracts. You white brothers have the faculty to burst the stoutest rocks. On our part we would not have disturbed those treaties. Shortly after our interview with our great Father, General Washington, at Philadelphia, a treaty was made at Canandaigua, by which we widened our former engagements with our white brothers, and made some new ones. The commissioner, Colonel Pickering, then told us that this treaty should be binding and should last, without alteration for two lives. We wished to make it extend much farther, and the Six Nations then wished to establish a lasting chain of friendship. On our part, we wished the treaty to last as long as trees grow, and waters run. Our Brother told us that he would agree to it.

"BROTHER: I have reminded you of what had taken place between our confederates, the Six Nations, and our white brothers, down to the treaty of Canandaigua. At the close of that treaty it was agreed, it being as strong and binding, as by my former comparisons I have explained, that if any difficulty should occur, if any monster should cross the chain of friendship, that we would unite to remove those difficulties, to drive away the monster; that we would go hand in hand and prolong the chain. So it was agreed.

"BROTHER: Many years ago we discovered a cloud rising that darkened the prospect of our peace and happiness. We heard eventful things from different quarters, from different persons, and at different times, and foresaw that the period was not very distant, when this threatening cloud would burst upon us.

"BROTHER: During the late war we intended to take no part. Yet residing within the limits of the United States, and with the advice of General Porter, we agreed around our council fire, that it was right, and we took a part. We thought it would help to promote our friendship with our white brothers, to aid the arms of the United States, and to make our present seats still stronger. These were our reasons. What were the results? We lost many of our warriors. We spilt our blood in a cause between you, and a people not of our color.

"BROTHER: These things may be new to you, but they are not new to your government. Records of these things are with our great Father, the President. You have come, therefore, for a very different purpose from the one we expected. You come to tell us of our situation, of our reservations, of the opinion of the President that we must change our old customs for new ones; that we must concentrate in order to enjoy the fair means you offer of civilization, and improvement in the arts of agriculture.

"BROTHER: At the treaty of Canandaigua, we were promised that different kinds of mechanics, blacksmiths, and carpenters, should be sent among us; and farmers with their families, that our women might learn to spin. We agreed to receive them. We even applied for these benefits. We were told that our children were too young to be taught. Neither farmers or mechanics were sent.

"BROTHER: We had thought that the promises made by one President, were handed down to the next. We do not change our chiefs as you do. Since these treaties were made, you have had several Presidents. We do not understand why the treaty made by one, is not binding on the other. On our part we expect to comply with our engagements.

"BROTHER: You told us when the country was surrounded by whites, and in possession of Indians, that it was unproductive, not being liable to taxes, nor to make roads nor improvements, it was time to change. As for the taxing of Indians, this is extra-ordinary; and was never heard of, since the settlement of America. The land is ours, by the gift of the Great Spirit. How can you tax it? We can make such roads as we want, and did so when the land was all ours. We are improving our condition. See these large stocks of cattle, and those fences. We are surrounded by the whites, from whom we can procure cattle, and whatever is necessary for our improvement. Now that we are confined to narrow limits, we can easily make our roads, and improve our lands.

"Look back to the first settlement by the whites, and then look at our present condition. Formerly we continued to grow in numbers, and in strength. What has become of the Indians, who extended to the salt water? They have been driven back and become few, while you have been growing numerous, and powerful. This lands is ours, from the God of Heaven. It was given to us. We cannot make land. Driven back and reduced as we are, you wish to cramp us more and more. You tell us of a pre-emptive right. Such men you say own one reservation, and such another. But they are all ours, ours from the top to the bottom. If Mr. Ogden had come from heaven, with flesh on his bones, as I we now see him, and said that the Heavenly Father had given him a title, we might then believe him.

"BROTHER: You say that the President has sent us word that it is for our interest to dispose of our lands. You tell us that there is a good tract of land at Allegany. This too is very extraordinary. Our feet have covered every inch of that reservation. A communication like this has never been made to us, at any of our councils. The President must have been disordered in mind, when he offered to lead us off by the arms, to the Allegany reservation. I have told you of the treaty we made with the United States. Here is the belt of wampum, that confirmed that treaty. Here too is the parchment. You know its contents. I will not open it. Now the tree of friendship is decaying; its limbs are fast falling off. You are at fault.

"Formerly we called the British brothers. Now we call the President, our Father. Probably among you, are persons with families of children. We consider ourselves the children of the President. What would be your feelings, were you told that your children were to be cast upon a naked rock, there to protect themselves? The different claims you tell us of, on our lands, I cannot understand. We are placed here by the Great Spirit, for purposes known to him. You have no right to interfere. You told us that we had large and unproductive tracts of land. We do not view it so. Our seats, we consider small; and if we are left here long, by the Great Spirit, we shall stand in need of them. We shall be in want of timber. Land after many years' use wears out; our fields must be renewed, and new ones improved, so that we have no more land in our reservations than we want. Look at the white people around us, and back. You are not cramped for lands. They are large. Look at that man. [Footnote: Mr. Ellicott, agent of the Holland Land Company.] If you want to buy, apply to him. He has lands enough to sell. We have none to part with. You laugh, but do not think I trifle. I am sincere. Do not think we are hasty in making up our minds. We have had many councils, and thought for a long time upon this subject. We will not part with any, not with one of our reservations.

"We recollect that Mr. Ogden addressed his speech to you, therefore I have spoken to you. Now I will speak to Mr. Ogden.

"BROTHER: You recollect when you first came to this ground, that you told us you had bought the pre-emptive right. A right to purchase given you by the government. Remember my reply. I told you, you had been unfortunate in buying. You said you would not disturb us. I then told you as long as I lived, you must not come forward to explain that right. You have come. See me before you. You have heard our reply to the commissioner sent by the President. I again repeat that, one and all, chiefs and warriors, we are of the same mind. We will not part with any of our reservations. Do not make your application anew, nor in any other shape. Let us hear no more of it. Let us part as we met, in friendship."

Col. Stone refers to the kindness of Major Joseph Delafield, for the speeches made at this council, as given in his work, and the most important of which is presented here; they were taken down at the time from the lips of the interpreter, who stated that "he could not translate some of Red Jacket's figurative flights, they were too wild and difficult to be rendered in English, and he did not attempt it." Much doubtless that served to give point and zest to his speech, was either omitted, or lost its force, in being transferred to our language. The writer of the sketch previously alluded to, among several points in this speech which were impressed on his memory, mentions one not found in the above. "The gentleman says, that our great Father says, we can go Allegany, and have a good foothold forever; *yes, a good foothold, for it is all rock.*"

Though the efforts of the Ogden Company to obtain the consent of the Indians to sell their remaining lands, were at this time unsuccessful, they were nevertheless repeated. The demand of Red Jacket, "do not make your application anew, nor in any other shape," was unheeded.

Col. Stone, on the authority of the Hon. Albert Tracy, mentions a treaty held for this same purpose in 1822 or 1823, in which Red Jacket replied to a speech made by the commissioner, and also by Governor Ogden, entering, as in the preceding speech, upon a regular and connected history of the transactions of the Indians with the whites, up to that time, and in the course of his speech, used the language very happily alluded to by Mr. Bryant, in his memorial address.

At the close of the speech that has been quoted almost entire, some of his people desired him to apologize for one or two utterances he had made, regarding them as rude, and adapted to awaken unpleasant reflections. He refused, saying, "NO, IT HAS GONE FORTH, LET IT STAND." A circumstance doubtless alluded to, in the words which immediately follow: "Often the fierceness of his temper, the righteous indignation that swelled his bosom, impelled him to hurl defiance at his foes, and to use language, the possible consequences of which, caused the more timid and abject of his followers, to tremble with apprehension. But Red Jacket would retract not a single word, although a majority of the chiefs, would sometimes secretly deprecate the severity of his utterances."

"Again on other occasions, sorely beset and almost despairing, he would essay to melt the hearts of the pitiless pursuers of his people, and give utterance to such touching words as these:

"We first knew you a feeble plant, which wanted a little earth whereon to grow. We gave it to you, and afterward, when we could have trod you under our feet, we watered and protected you; and now you have grown to be a mighty tree, whose top reaches the clouds, and whose branches overspread the whole land; whilst we, who were then the tall pine of the forest, have become the feeble plant and need your protection."

"Again assuming the pleading tones of a supplicant, he said, 'when you first came here, you clung around our knee, and called us FATHER. We took you by the hand and called you BROTHERS. You have grown greater than we, so that we can no longer reach up to your hand. But we wish to cling around your knee, and be called YOUR CHILDREN.'"

In this same speech, referring to their services during the late war with England, he said:

"Not long ago you raised the war-club against him, who was once our great Father over the waters. You asked us to go with you to the war. It was not our quarrel. We knew not that you were right. We asked not; we cared not; it was enough for us, that you were our brothers. We went with you to the battle. We fought and bled for you; and now," his eye kindling with emotion, and the deepest feeling indicated in his utterance, as he pointed to some Indians present, that had been wounded in that contest; "and now, dare you pretend to us, that our Father the President, while he sees our blood running, yet fresh from the wounds received, while fighting his battles, has sent you with a message to persuade us to relinquish the poor remains of our once boundless possessions; to sell the birth place of our children, and the graves of our fathers! No! Sooner than believe that he gave you this message, we will believe that *you have stolen your commission, and are a cheat and a liar*."

Once more, speaking of the pre-emptive right and the assurance given them that their lands were desired only in return for a fair equivalent of their value, he called their attention to the great cessions the Indians had already made, together with the solemn declarations that they should not be importuned to relinquish their remaining reservations, he said: "You tell us of your claim to our land, and that you have purchased it from your State. We know nothing of your claim, and we care nothing for it. Even the whites have a law by which they cannot sell what they do not own. How then has your State, which never owned our land, sold it to you? We have a title to it, and we know that our title is good; for it came direct from the Great Spirit, who gave it to us, his red children. When you can ascend where he is," pointing toward the skies, "and will get his deed, and show it to us, then, and never till then, will we acknowledge your title. You say you came not to cheat us of our lands, but to buy them. Who told you that we have lands to sell? You never heard it from us."

Then rising up and giving Mr. Ogden a look of deep earnestness, if not of indignation, he said:

"Did I not tell you the last time we met, that whilst Red Jacket lived, you would get no more lands of the Indians? How then, while you see him alive and strong," striking his hand violently on his breast, *"do you think to make him a liar?"*

The persistence with which the Senecas were importuned to sell their lands, led them to make an appeal to the president, and afterward to the governor of New York.

The latter, Governor De Witt Clinton, sent them a reply worthy of his name and office. It is as follows:

"All the right that Ogden and his company have to your reservations, is the right of purchasing them when you think it expedient to sell them, that is, they can buy your lands, but no other person can. You may retain them as long as you please, and you may sell them to Ogden as soon as you please. You are the owners of these lands in the same way that your brethren the Oneidas, are of their reservations. They are all that is left of what the Great Spirit gave to your ancestors. No man shall deprive you of them without your consent. The State will protect you in the full enjoyment of your property. We are strong and willing to shield you from oppression. The Great Spirit looks down on the conduct of mankind, and will punish us if we permit the remnant of the Indian nations which is with us to be injured. We feel for you, brethren; we shall watch over your interests. We know that in a future world we shall be called upon to answer for our conduct to our fellow creatures."

Col. Stone refers to the Hon. Albert H. Tracy, as having furnished the notes of the council we have just been considering. The same authority speaking of the eloquence of Red Jacket, says: "It is evident that the best translations of Indian speeches, must fail to express the beauty and sublimity of the originals; especially of such an original as Red Jacket. It has been my good fortune to hear him a few times, but only of late years, and when his powers were enfeebled by age, and still more, by intemperance. But I shall never forget the impression made on me, the first time I saw him in council:

"Deep on his front engraven,
"Deliberation sate, and public care,
"And princely counsel in his face yet shone,
"Majestic, though in ruin.

"I can give no idea of the strong impression it made on my mind, though conveyed to it through the medium of an illiterate interpreter, Even in this mangled form, I saw the *disjecta membra* of a regular and splendid oration."[Footnote: Col. Stone's Life and Times of Red Jacket.]

The Ogden Company though defeated time and again by the watchfulness, and powerful influence of Red Jacket, continued to ply their endeavors, until by degrees, the remaining portion of their once proud inheritance, was wrested from them, and the orator was left in the decline of life to survey, as he often did in a spirit of dejection, the haunts of his youth, which had nearly all passed into other hands, through the craft and avarice of the white man.

CHAPTER XIX

Witchcraft—Lease of Tom-Jemmy—Testimony of Red Jacket—Red Jacket's Philippic—Finding of the court—Remarkable interview of Dr. Breckenridge with Red Jacket—Further expression of views.

In the spring of 1821, a man belonging to Red Jacket's tribe, fell into a languishing condition, and after lingering for some time, unable to obtain relief, died. The *medicine men* were unable to divine the cause of his malady; the circumstances of his sickness and death, were thought to be very peculiar, and his friends could discover no better way of explaining the matter, than to suppose he had been bewitched.

The Indians believed in sorcery, and at different times in their history had been known to execute summary judgment, on those whom they supposed to be guilty of practicing the Satanic art. In the present instance suspicion rested on the woman, by whom he had been attended, during his sickness. In pursuance of the customs of their nation she was condemned to die. The sentence was executed by Soo-nong-gise, a chief, commonly called Tom- Jemmy. It took place at their reservation near Buffalo. Coming to the knowledge of the whites in the vicinity, it excited feelings of horror, mingled with indignation. The case was taken in hand by their authorities, who without regard to Indian jurisdiction, arrested Tom-Jemmy and threw him into prison.

At his trial the plea was set up in his defense, that the Indians were a sovereign and independent nation, having their own laws, and their own mode of carrying them into execution; that the offense was within the acknowledged bounds of their own territory, that according to their laws, it was not a crime, inasmuch as the act of the prisoner was in the execution of a sentence, that had been passed upon the woman in question.

The trial was conducted with reference to this issue, and numerous witnesses were examined to substantiate the facts having a bearing on the case. Red Jacket, among others, was called upon the stand, and examined with reference to the laws, and usages of his people.

The counsel who conducted the prosecution, wishing to exclude his testimony, inquired whether he believed in the existence of a God? *"More truly than one who could ask me such a question;"* was his instant and indignant reply.

On cross examination the inquiry was made, as to the rank he held among his own people? "Look at the papers, which the white men keep the most carefully," meaning the treaties ceding their lands, "and they will tell you."

The orator's testimony, as did also that of other witnesses, who testified in the case, went to show that this woman, according to the judgment of the Indians, was a witch. That she had been regularly tried, and condemned by their laws; and her death was in conformity with usages, that had been in existence among them, from time immemorial.

During the course of this examination, Red Jacket perceived that the belief of the Indians in witchcraft, was made a subject of ridicule among the bystanders, as well as legal gentlemen present, and he took occasion when an opportunity offered, to break forth in the following language:

"What! Do you denounce us as fools and bigots, because we still believe that which you yourselves believed two centuries ago? Your black coats thundered this doctrine from the pulpit, your judges pronounced it from the bench, and sanctioned it with the formalities of law; and you would now punish our unfortunate brother, for adhering to the faith of *his* fathers and of *yours*! Go to Salem! Look at the records of your own government, and you will find that hundreds have been executed for the very crime, which has called forth the sentence of condemnation against this woman, and drawn down upon her the arms of vengeance. What have our brothers done, more than the rulers of your own people have done? And what crime has this man committed, by executing in a summary way, the laws of his country, and the command of the Great Spirit?" [Footnote: Col. Stone, and also Drake's Book of the Indians.]

It has been observed of Red Jacket's appearance on this occasion: "there is not, perhaps in nature, a more expressive eye than that of Red Jacket; when fired by indignation or revenge, it is terrible; and when he chooses to display his unrivalled talent for irony, his keen sarcastic glance, is irresistible." [Footnote: Drake.]

This trial resulted in finding the allegations in the prisoner's plea to be true; yet the judgment being suspended, it was referred finally to the Supreme Court. A thorough examination of the laws, treaties and history relating to our correspondence with the Indian tribes, gave evidence of a sort of sovereignty among them, but as it was thought inexpedient to render a decision, that would recognize their independent jurisdiction, the prisoner was liberated, and the case dismissed.

Not far from the time we are now considering, a remarkable conversation took place between Red Jacket and a young candidate for the clerical office, who afterward became an eminent divine. [Footnote: Rev. John Breckenridge, D. D.] It serves very much to illustrate the orator's character and views, and as we have permission, we give it entire, as follows:

"The first-opportunity I ever enjoyed of seeing that deservedly celebrated Indian chief, Red Jacket, was in the year 1821, at the residence of General Peter B. Porter, Black Rock, New York. Being on a visit to the general and his family, it seemed a peculiarly fit occasion to become acquainted with the great Seneca orator, whose tribe resided within a few miles of Black Rock. General Porter embraced in his command, the Indian warriors who fought with us on that line, during the late war, with Great Britain. From this cause; from his high character; his intimate acquaintance with the chiefs; and his known attachment to these interesting people, he had great influence over them; and his lamented lady, who it is not indelicate for me to say, was my sister, had by her kindness won the rugged hearts of all their leading men. So that their united influence, and my near relationship to them, secured to me at once access to the chiefs, and their entire confidence.

"I had not only a great desire to see Red Jacket, but also to use this important opportunity to correct some of his false impressions, in regard to Christianity, and the missionaries established in his tribe. To this end it was agreed to invite Red Jacket and the other chiefs of the Senecas, to visit Co-na-shus-ta, [Footnote: Name given by Red Jacket to General Porter.] and meet his brother at his house. The invitation was accordingly given, and very promptly and respectfully accepted.

"On the appointed day they made their appearance in due form headed by Red Jacket, to the number of perhaps eight or ten, besides himself. Red Jacket was dressed with much taste, in the Indian costume throughout. He wore a blue dress, the upper garment cut after the fashion of a hunting shirt, with blue leggings, very neat moccasins, a red jacket, and a girdle of red about his waist. I have seldom seen a more dignified or noble looking body of men than the entire group. It seems, though no such impression was designed to be made by the terms of the invitation, that some indefinite expectation had been excited in their minds, of meeting an official agent on important business. And they have been so unworthily tampered with, and so badly treated by us, as a people, and many of their most important treaties have been so much the result of private and corrupting appeals, that they very naturally look for some evil design in every approach to them, however open and simple it may be. So it was on this occasion. As soon as the ceremonies of introduction had passed, with the civilities growing out of it, the old orator seated himself in the midst of the circle of chiefs, and after a word with them, followed by a general assent, he proceeded in a very serious and commanding manner, always speaking in his own nervous tongue, through an interpreter, to address me as follows:

"'We have had a call from our good friends,' (pointing to the general and his lady), 'to come down to Black Rock to meet their brother. We are glad to break bread and to drink the cup of friendship with them. They are great friends to our people, and we love them much. Co-na-shus-ta is a great man. His woman has none like her. We often come to their house. We thank them for telling us to come to-day. But as all the chiefs were asked we expected some important talk. Now, here we are: what is your business?'

"This, as may be readily supposed, was an embarrassing position to a young man just out of college. I paused. Every countenance was fixed upon me, while Red Jacket in particular seemed to search me with his arrowy eye, and to feel that the private and informal nature of the meeting, and the extreme youth of the man, were hardly in keeping with the character and number of the guests invited; and his whole manner implied, that 'but for the sake of the general and his good viands, I should have waited for you to come to us.' With these impressions of his feelings, I proceeded to say in reply:

"That I should have thought it very presumptuous in me to send for him alone, and still more for all the chiefs of his tribe, to come so far to see me; and that my intention had been to visit him, and the other chiefs at his town; but the general and his lady, could not go with me to introduce me. Nor were we at all certain that we should find him and the other chiefs at home; and at any rate the general's house was more convenient. He intended, when he asked them, to keep them as long as they could stay, and to invite them to break his bread, and drink his cup, and smoke his pipe; that his woman, and he as well as I, desired to see them at their house; that as to myself, I was a young man, and had no business with them, except that I had heard a great deal of Red Jacket, and wished to see him and hear him talk; and also that I had some things to say to him, when we were better acquainted, which though not *business*, were important to his people; and I thought it would be interesting to him, as I knew he loved his people much; and finally that I would return his visit, and show him that it was not out of disrespect, but out of regard for him, and great desire to see him, that we had sent for him, this being the way that white men honor one another.

"Mrs. Porter immediately confirmed what I had said, and gave special point to the hospitality of the house, and the great desire I had to see Red Jacket. Her appeal, added to the reply, relaxed the rigor of his manner and that of the other chiefs, while it relieved our interview of all painful feelings.

"After this general letting down of the scene, Red Jacket turned to me familiarly and asked; 'What are you? You say you are not a government agent, are you a gambler? [Footnote: The name given by Red Jacket to a land speculator.] or a black coat? or what are you?' I answered: 'I am yet too young a man to engage in any profession: but I hope some of these days to be a black coat.' He lifted up his hands accompanied by his eyes, in a most expressive way, and though not a word was uttered, every one fully understood that he very distinctly expressed the sentiment, what a fool!

"I had too often been called to bear from those reputed great and wise among *white* men, the shame of the cross, to be surprised by his manner; and I was too anxious to conciliate his good feelings to attempt any retort, so that I commanded my countenance, and seeming not to have observed him, I proceeded to tell him something about our colleges, etc., etc. That gradually led his mind away from the ideas with which it was filled and excited when he arrived.

"A good deal of general conversation ensued, addressed to one and another of the chiefs, and we were just arriving at the hour of dinner, when our conference was suddenly broken up by the arrival of a breathless messenger, saying that an old chief, whose name I forget, had just died, and the other chiefs were immediately needed to attend his burial. One of the chiefs shed tears at the news; all seemed serious; but the others suppressed their feelings, and spent a few moments in very earnest conversation, the result of which Red Jacket announced to us. They had determined to return at once to their village; but consented to leave Red Jacket and his interpreter. In vain were they urged to wait until after dinner, or to refresh themselves with something eaten by the way. With hurried farewell and quick steps they left the house, and by the nearest footpath returned home.

"This occurrence relieved me of one difficulty. It enabled me to see Red Jacket at leisure and alone. It seemed also to soften his feelings, and make him more affable and kind.

"Soon after the departure of the chiefs, we were ushered to dinner. Red Jacket behaved with great propriety, in all respects; his interpreter, Major Berry, though half a white man and perhaps a chief, eat like a true savage. After a few awkward attempts at the knife and fork, he found himself falling behind, and repeating the old adage which is often quoted to cover the same style among our white urchins of picking a chicken-bone, '*that fingers were made before knives and forks*,' he proceeded with real gusto, and much good humor, to make up his lost time upon all parts of the dinner. It being over, I invited Red Jacket into the general's office, where we had, for four hours a most interesting conversation on a variety of topics, but chiefly connected with Christianity; the government of the United States; the missionaries; and his loved lands.

"So great a length of time has passed since that interview, that there must be supposed a failure in the attempt perfectly to report what was said. I am well assured I cannot do justice to his language, even as diluted by the ignorant interpreter; and his manner cannot be described. But it was so impressive a conversation, and I have so often been called on to repeat it, that the substance of his remarks has been faithfully retained by my memory. It is only attempted here to recite a small part of what was then said, and that with particular reference to the illustration of his character, mind and opinions.

"It has already been mentioned and is largely known, that Red Jacket cherished the most violent antipathy toward the American missionaries, who had been located among his people. This led to very strenuous resistance of their influence, and to hatred of their religion, but of the true character of which, he was totally ignorant. His deep attachment to his people, and his great principle that their national glory and even existence, depended upon keeping themselves distinct from white men, lay at the foundation of his aversion to Christianity. Though a pagan, yet his opposition was political, and he cared very little for any religion except so far as it seemed to advance, or endanger the glory and safety of the tribe.

"He had unfortunately been led by designing and corrupt white men, who were interested in the result, falsely to associate the labors of the missionaries, with designs against his nation; and those who wished the Senecas removed from their lands that *they* might profit by the purchase, and who saw in the success of the mission the chief danger to *their* plans, artfully enlisted the pagan party, of which Red Jacket was the leader, to oppose the missionaries, and thus effectually led to the final frustration of Red Jacket's policy; in and by the defeat of the missionary enterprise. But as this question is discussed in the sequel, I will not anticipate. Thus much it was necessary to premise, in order to explain the nature and ends of my interview with Red Jacket.

"My object was to explain the true state of the case to him, and after this to recommend the doctrine of Christ to his understanding and heart. My first step, therefore, was to ask him why he so strongly opposed the settlement and labors of the missionaries? He replied, because they are the enemies of the Indians, and under the cloak of doing them good are trying to cheat them out of their lands. I asked him what proof he had of this. He said he had been told so by some of his wise and good friends, among the white men, and he observed that the missionaries were constantly wanting more land, and that by little and little, for themselves, or those who hired them to do it, they would take away all their lands, and drive them off.

"I asked him if he knew there was a body of white men, who had already bought the exclusive right to buy their lands, from the government of New York, and that therefore the missionaries could not hold the lands given or sold them by the Indians, a moment after the latter left their lands and went away. He seemed to be startled by the statement, but said nothing. I proceeded to tell him that the true effect of the missionary influence on the tribe was to secure to them the possession of their lands, by civilizing them, and making them quit the chase, for the cultivation of the soil, building good houses, educating their children, and making them permanent citizens and good men. This was what the speculators did not wish. Therefore they hated the missionaries. He acknowledged that the Christian party among the Indians did as I said; but that was not the way for an Indian to do. Hunting, war and manly pursuits, were best fitted to them. But, said I, your reservation of land is too little for that purpose. It is surrounded by the white people, like a small island by the sea; the deer, the buffalo and bear, have all gone. This won't do. If you intend to live so much longer, you will have to go to the great western wilderness, where there is plenty of game, and no white men to trouble you. But he said, we wish to keep our lands and to be buried by our fathers. I know it, and therefore I say that the missionaries are your best friends; for if you follow the ways they teach, you can still hold your lands, though you cannot have hunting grounds, and therefore you must either do like white men, or remove from your lands, very soon. Your plan of keeping the Indians distinct from the white people is begun too late. If you would do it and have large grounds, and would let the missionaries teach you Christianity, far from the bad habits and big farms of the white people, it would then be well: it would keep your people from being corrupted, and swallowed up by our people who grow so fast around you, and many of whom are very bad. But it is too late to do it here, and you must choose between keeping the missionaries, and being like white men, and going to a far country: as it is, I continued, Red Jacket is doing more than any body else to break up and drive away his people.

"This conversation had much effect upon him. He grasped my hand and said if that were the case it was new to him. He also said he would lay it up in his mind (putting his hand to his noble forehead), and talk of it to the chiefs, and the people.

"It is a very striking fact that the disgraceful scenes now passing before the public eye over the grave of Red Jacket, so early and so sadly fulfil these predictions; and I cannot here forbear to add that the thanks of the nation are due to our present chief-magistrate, [Footnote: The President alluded to is Mr. Van Buren.—W. L. S.] for the firmness with which he has resisted the recent efforts to force a fraudulent treaty on the remnant of this injured people, and drive them against their will, and against law and treaties sacredly made, away from their lands, to satisfy the rapacity of unprincipled men.

"It may be proper here to say likewise, that I do by no means intend to justify, all that possibly may have been done by the missionaries to the Senecas. It is probable the earliest efforts were badly conducted; and men of more ability ought to have been sent to that peculiar and difficult station. But it is not for a moment to be admitted, nor is it credible that the authors of the charge believe it, that the worthy men who at every sacrifice went to the mission among the Senecas, had any other than the purest purposes. I visited the station, and intimately knew the chief missionary. I marked carefully their plan and progress, and do not doubt their usefulness any more than their uprightness; and beyond all doubt it was owing chiefly to malignant influence exerted by white men, that they finally failed in their benevolent designs. But my business is to narrate, not to discuss.

"My next object was to talk with Red Jacket about Christianity itself. He was prompt in his replies, and exercised and encouraged frankness, with a spirit becoming a great man.

"He admitted both its truth and excellence, as adapted to white men. He said some keenly sarcastic things about the treatment that so good a man as Jesus, had received from white men. The white men, he said, ought all to be sent to hell for killing him; but as the Indians had no hand in that transaction, they were in that matter innocent. Jesus Christ was not sent to them; the atonement was not made for them; nor the Bible given to them; and therefore the Christian religion, was not meant for them. If the Great Spirit had intended that the Indians should be Christians, he would have made his revelation to them, as well as to the white men. Not having done so, it was clearly his will that they should continue in the faith of their fathers. He said that the red man was of a totally different race, and needed an entirely different religion, and that it was idle as well as unkind, to try to alter their religion, and give them ours.

"I asked him to point out the difference of the races, contending that they were one, and needed but one religion, and that Christianity was that religion, which Christ intended for, and ordered to be preached, to all men. He had no distinct views of the nature of Christianity as a method of salvation, and denied the need of it. As to the unity of the races, I asked if he ever knew two distinct races, even of the lower animals to propagate their seed from generation to generation. But do not Indians and white men do so? He allowed it; but denied that it proved the matter in hand. I pressed the points of resemblance in every thing but color, and that in the case of the Christian Indians there was a common mind on religion. He finally waived this part of the debate, by saying that one thing was certain, whatever else was not, that white men had a great love for Indian women, and left their traces behind them wherever they could!

"On the point of needing pardon, from being wicked, he said the Indians were *good* till the white man corrupted them. But did not the Indians have *some* wickedness *before* that? 'Not so much.' And how was *that* regarded by the Great Spirit?—Would he forgive it? He hoped so, 'did not know.' Jesus, I rejoined, came to tell us He would, and to get that pardon for us.

"As to suffering and death among the Indians, did not they prove that the Great Spirit was angry with *them*, as well as with white men? Would he thus treat men that were *good*? He said they were not wicked before white men came to their country, and taught them to be so. But they *died before that*? And why did they die, if the Great Spirit was not angry, and they wicked? He could not say, and in reply to my explanation of the gospel doctrine of the entrance of death by sin, he again turned the subject by saying he was a 'great doctor,' and could cure any thing but death.

"The interpreter had incidentally mentioned that the reason the chiefs had to go home so soon, was that they always *sacrificed a white dog on the death of a great man*. I turned this fact to the account of the argument, and endeavored to connect it with, and explain by it the doctrine of atonement, by the blood of Christ, and also pressed him on the questions, how can this *please* the Great Spirit on *your* plan? Why do you offer such a *sacrifice*, for so it is considered? And *where* they got such a rite from? He attempted no definite reply. Many other topics were talked over. But these specimens suffice to illustrate his views, and mode of thinking.

"At the close of the conversation he proposed giving me a *name*, that henceforth I might be numbered among his friends, and admitted to the intercourse and regards of the nation. Supposing this not amiss, I consented. But before he proceeded he called for some whiskey. He was at this time an intemperate man, and though perfectly sober on that occasion, evidently displayed toward the close of the interview, the need of stimulus, which it is hardly necessary to say, we carefully kept from him. But he *insisted* now, and after some time a small portion was sent to him in the bottom of a decanter. He looked at it, shook it, and with a sneer said, 'why here is not whiskey enough for a name to float in.' But no movement being made to get more, he drank it off, and proceeded with a sort of pagan orgies, to give me a name. It seemed a semi-civil, semi- religious ceremony. He walked around me again and again, muttering sounds which the interpreter did not venture to explain; and laying his hand on me pronounced me 'Con-go-gu-wah,' and instantly, with great apparent delight, took me by the hand as a brother. I felt badly during the scene, but it was beyond recall, and supposing it might be useful in a future day, submitted to the initiation.

"Red Jacket was in appearance nearly sixty years old at this time. He had a weather-beaten look; age had done something to produce this, probably intemperance more. But still his general appearance was striking, and his face noble. His lofty and capacious forehead, his piercing black eye, his gently curved lips, and slightly aquiline nose, all marked a great man, and as sustained and expressed by his dignified air, made a deep impression on every one that saw him. All these features became doubly expressive when his mind and body were set in motion by the effort of speaking, if effort that may be called which flowed like a free, full stream from his lips. I saw him in the wane of life, and I heard him only in private, and through a stupid, careless interpreter. Yet notwithstanding these disadvantages, he was one of the greatest men and most eloquent orators I ever knew. His cadence was measured and yet very musical. In ordinary utterance it amounted to a sort of musical monotony. But when excited he would spring to his feet, elevate his head, expand his arms and utter with indescribable effect of manner and tone, some of his noblest thoughts.

"After this interesting conference had closed, the old chief with his interpreter, bade us a very civil and kind farewell, and set forth on foot for his own wigwam.

"It was four years after this before I had the pleasure of again seeing my old friend. I was then on a flying visit to Black Rock. At an early day I repaired to his village, but he was not at home. Ten days after, as we were just leaving the shore in the steamboat to go up the lake, he suddenly presented himself. It was unhappily too late to return. He hailed me by name, and pointed with much animation to such parts of his person as were decorated with some red cloth which I at parting had presented to him, and which, though not worn as a jacket, was with much taste distributed over his person. These he exhibited as proofs of his friendly recollection.

"The last time I ever saw him was at the close of Mr. Adams' administration. He, with a new interpreter (Major Berry having been removed by death), had been on a visit to his old friend, Co-na-shus-tah, then Secretary of War. After spending some time at the capital, where I often met him, and had the horror to see his dignity often laid in the dust, by excessive drunkenness, he paid me by invitation a final visit at Baltimore, on his way home. He took only time enough to dine. He looked dejected and forlorn. He and his interpreter had each a suit of common infantry uniform, and a sword as common, which he said had been presented to him at the war department. He was evidently ashamed of them. I confess I was too. But I forbear. He was then sober and serious. He drank hard cider, which was the strongest drink I could conscientiously offer him, so I told him. He said it was enough. I said but little to him of religion, urged him to prepare to meet the Great Spirit, and recommended him to go to Jesus for all he needed. He took it kindly, said he should see me no more, and was going to his people to die. So it was, not long after this, he was called to his last account."

Col. Stone represents the testimony of Dr. Breckenridge as corresponding with hundreds of others, who confess their inability to do the orator justice. He laments "his inability to make even an approach to justice, as to the language, and figures in which Red Jacket clothed his thoughts, and by which he illustrated and enforced them."

At another time the benefits of Christianity and the advantages of civilization, being urged by a benevolent gentleman on Red Jacket's attention, he made use of the following language: "As to civilization among the white people, I believe it is a good thing, and that it was so ordered they should get their living in that manner. I believe in a God, and that it was ordered by Him that we, the red people, should get our living in a different way, namely: from the wild game of the woods, and the fishes of the waters. I believe in the Great Spirit who created the heavens and the earth. He peopled the forests, and the air and the waters. He then created man and placed him as the superior animal of this creation, and designed him as governor over all other created beings on earth. He created man differing from all other animals. He created the red man, the white, the black and the yellow. All these he created for wise but inscrutable purposes."

Reasoning from analogy and from the different varieties of the same species, and the different species under one genus, among all other animals, he pointed out their different modes of living, and the different designs of the Creator, that appeared to be evinced with respect to them. He then proceeded:

"This being so, what proof have we that he did not make a similar arrangement with the human species, when we find so vast, so various, and so irreconcilable a variety among them, causing them to live differently, and to pursue different occupations.

"As to religion, we all ought to have it. We should adore and worship our Creator, for his great favor in placing us over all his works. If we cannot with the same fluency of speech, and in the same flowing language, worship as you do, we have our mode of adoring, which we do with a sincere heart; then can you say that our prayers and thanksgivings, proceeding from grateful hearts, and sincere minds are less acceptable to the Great God of the heavens and the earth, though manifested either by speaking, dancing, or feasting, than yours, uttered in your own manner and style?" [Footnote: As quoted by Col. Stone from MS collections of Joseph W. Moulton.]

CHAPTER XX

Personal characteristics—Interview with General Lafayette—Visit of a French Nobleman—Col. Pickering reproved—Address on launching a schooner bearing his name—Anecdote of Red Jacket and Capt. Jones—His humor— Strong memory—Its cultivation—Contempt for pretension without merit— Love for the sublime—Portraits—Acute perception—Refined sense of propriety—First bridge at Niagara Falls—Loss of his children—Care for his people.

A prominent characteristic of Red Jacket's mind, was self esteem, which led him to be quite tenacious of his own opinion. He probably did not underrate his own ability. He felt conscious of possessing talents, which would enable him to act with dignity and propriety, in any emergency calling for their exercise. He never appeared to be intimidated or embarrassed at the thought of meeting with great men, but seemed always to be at home in their society, and to feel and act as though he regarded himself on an equality with them. This was evident in his interview with General Lafayette, in 1825.

On being presented to the general, the orator inquired if he recollected being present, at the treaty of peace with the Six Nations at Fort Stanwix, in 1784. Lafayette replied that he remembered that great council very well. "And what," said he, "has become of the young chief, who resisted so strenuously and eloquently on that occasion, the idea of the Indians' burying the hatchet?"

"He is before you," was the instant reply. Upon which the general remarked, that time had wrought very great changes upon them both since that memorable period. "Ah!" said Red Jacket, "time has not been so severe on you, as it has on me. It has left you a fresh countenance, and hair to cover your head; while to me,—behold!"—And taking a handkerchief from his head, with an air of much feeling, he disclosed the fact that he was nearly bald. Several persons present could not refrain from smiling at the simplicity of the Indian, who appeared ignorant of the way the white man, was wont to repair the ravages of age in this respect. His simplicity was enlightened by the fact, that the general was indebted to a wig, for his generous supply of hair. Whereupon the orator playfully remarked, referring to the practice

of his people in war, that it had not occurred to him before, that he might supply the deficiency by *scalping* some of his neighbors. M. Lavasseur, the secretary of General Lafayette, remarks of the orator's appearance at that time. "This extraordinary man, although much worn down by time and intemperance, preserves yet in a surprising degree, the exercise of all his faculties. He obstinately refuses to speak any language, but that of his own people, and affects a great dislike to all others. Although it is easy to discern, that he perfectly understands the English. He refused nevertheless, to reply to the general before his interpreter had translated his questions into the Seneca language." [Footnote: See Drake, Col. Stone and others.]

A few Indian words, which the general had picked up during his previous visit to this country, on being repeated by him to the orator, gratified him exceedingly, and appeared to increase very much his regard for Lafayette.

Red Jacket appeared always to be gratified by attentions received from distinguished characters. Yet even to enjoy their society, he would not compromise his own dignity. It is said that "about the year 1820, a young French nobleman, who was making the tour of the United States, visited the town of Buffalo. Hearing of the fame of Red Jacket, and learning that his residence was but seven miles distant, he sent him word, that he was desirous to see him, adding a request that the chief would visit him in Buffalo the next day. Red Jacket received the message with contempt, and replied: 'Tell the *young man* that if he wishes to see the *old chief*, he may find him with his nation, where other strangers pay their respects to him; and Red Jacket will be glad to see him.'

"The count sent back his mesenger to say he was fatigued with his journey, and could not go to the Seneca village; that he had come all the way from France, to see the great orator of the Senecas, and after having put himself to so much trouble, to see so distinguished a man, the latter could not refuse to meet him at Buffalo.

"'Tell him,' said the sarcastic chief, 'It is very strange he should come so far to see me, and then stop within seven miles of my lodge.' The retort was richly merited. The count visited him at his wigwam, and then Red Jacket accepted an invitation to dine with him, at his lodgings at Buffalo.

"The young nobleman was greatly pleased with him, declaring that he considered him a greater wonder than the falls of Niagara. This remark was the more striking as it was made within view of the great cataract. But it was just. He who made the world, and filled it with wonders, has declared man to be the crowning work of the whole Creation." [Footnote: McKenney's Indian Biography.]

On one occasion at a treaty attended by Colonel Pickering, Red Jacket observed that the attention of the colonel, who was in the habit of taking down, as they were interpreted, the Indian speeches made, was withdrawn from himself, and his eye directed to the paper on which he was writing. Red Jacket paused. The colonel desired him to proceed. "No," said the orator, "not when you hold down your head." "Why can you not go on while I write?" "Because," replied the chief, "if you look me in the eye, you will then perceive if I tell you the truth or not." [Footnote: Col. Stone.]

On another occasion, Colonel Pickering turned, while the orator was addressing him to speak to a person near. The chief thereupon rebuked him, saying with much emphasis, "When a Seneca speaks he ought to be listened to with attention, from one extremity of this great island to the other." [Footnote: Ib.]

Toward the close of his life he was present by invitation, at the launching of a schooner at Black Rock, bearing his name. He made a short address on the occasion which indicates the estimation in which he regarded his own merit. In the course of his speech, addressing himself directly to the vessel, he said: "You have a great name given you, strive to deserve it. Be brave and daring. Go boldly into the great lakes, and fear neither the swift winds, nor the strong waves. Be not frightened nor overcome by them, for it is by resisting storms and tempests, that I, whose name you bear, obtained my renown. Let my great example inspire you to courage, and lead you to glory." [Footnote: Col. Stone.]

Also late in life, when at one of the hotels in Auburn, N. Y., observing a person whom he thought did not treat him with proper deference, he came and stood before him and stamping his foot on the floor, exclaimed with much emphasis, "*I am Red Jacket!*" [Footnote: Incident given to the author by J. C. Ivison, Esq., of Auburn.]

He did not relish being trifled with even in playfulness.

"At one time when visiting the house of Captain Jones, on taking his seat at the breakfast table with the family, Mrs. Jones, knowing his extreme fondness for sugar, mischieviously prepared his coffee without the addition of that luxury. On discovering the cheat, the chief looked at the captain with an offended expression, and thus rebuked him: 'My son,' stirring his cup with energy, 'Do you allow your squaw thus to trifle with your father?' Perceiving at the same time, by the giggling of the children, that they had entered into the joke, he continued, 'And do you allow your children to make sport of their chief?' Jones and his wife thereupon apologized, and the latter made the *amende honorable*, by handing him the sugar-bowl, which he took, and with half angry sarcasm filled the cup to the brim, with sugar.

The liquid not holding so large a quantity in solution, he ate the whole with his spoon." [Footnote: Col. Stone.]

Still he enjoyed a laugh when he was making the sport. He was very entertaining in conversation, and would sometimes in the presence of his associates, relax his dignity, and for a time, when he felt in the mood, keep them in a roar of laughter, by his anecdotes, or by taking off something ludicrous, he had observed among the whites. When he had carried it sufficiently far, he would draw himself up, and resume his dignity, when by common consent, the sport would cease. [Footnote: Wm. Jones, to the author.]

He very often entertained his people also, by recounting his interviews with distinguished persons, or by describing what he had seen in great places.

One conversant with him thus speaks of the manner in which he represented to his people, what he had seen during his visit at the seat of government. "I remember having seen him on one of those occasions, when, after having seated the Indians around him in a semi-circle, taking the cocked hat that had been presented to him by General Knox, then Secretary of War, in his hand, he went round bowing to the Indians, as though they were the company at the president's house, and himself the president. He would then repeat to one and another all the compliments which he chose to suppose the president had bestowed upon him, and which his auditors and admiring people, supposed had been thus bestowed." [Footnote: Thomas Morris to Col. Stone.]

Red Jacket had a very *tenacious memory*. The Indians were noted for the care they bestowed on this faculty of the mind. In the absence of written records, they formed a device, which was quite ingenious, and indicated a high degree of intelligence, by which they perpetuated the knowledge of important events, in their history. They used belts, and strings of wampum.

For instance, they are assembled to form some important treaty. This *treaty* would be represented by the *belt*. Each string in that belt would represent a distinct article, or provision in that treaty. As they fixed their eye upon the belt, they knew it as well as though it had been labelled. As they took hold of each string, they could as it were, read each article of the treaty. For the preservation of these belts they had what were termed their council-houses, where they were hung up in order, and preserved with great care. At times they were reviewed. The father would go over them, and tell the meaning of each belt and of each string in the belt to the son, and thus the knowledge of all their important events, was transmitted from one generation to another.

Red Jacket, without any doubt excelled all of his race, in the perfection to which he had brought this faculty of his mind. Nothing escaped the tenacious grasp of his memory.

The following is an instance in point. At a council held with the Indians by Gov. Tompkins of New York, a contest arose between him and Red Jacket in regard to a fact connected with a treaty of many years' standing. Mr. Tompkins stated one thing, and the Indian chief corrected him, insisting that the reverse of his assertion was true. "But" it was rejoined: "you have forgotten." We have it written down on paper. "The paper then tells a lie," was the confident answer; "I have it written down here;" he added, placing his hand with great dignity on his brow. "You Yankees are born with a feather between your fingers, but your paper does not speak the truth. The Indian keeps his knowledge here. This is the book the Great Spirit gave them; it does not lie." A reference was immediately made to the treaty in question, when to the astonishment of all present, and the triumph of the unlettered statesman, the document confirmed every word he had uttered. [Footnote: McKenney's Indian Biography.]

He held in utter contempt *pretensions* without *merit*. "On one occasion not many years before his death, a gentleman from Albany, on a visit at Buffalo, being desirous of seeing the chief, sent a message to that effect. The gentleman was affluent in money and in words, the latter flowing forth with great rapidity, and in an inverse ratio to his ideas. He had also a habit of approaching very near to any person with whom he was conversing, and chattering with almost unapproachable volubility. On receiving the message, Red Jacket dressed himself with the utmost care, designing, as he ever did when sober, to make the most imposing impression, and came over to the village.

"Being introduced to the stranger, he soon measured his intellectual capacity, and made no effort to suppress his disappointment, which was indeed sufficiently disclosed in his features. After listening, for a few moments to the chatter of the gentleman, Red Jacket with a look of mingled chagrin and contempt, approached close to him and exclaimed, 'cha, cha, cha,' as rapidly as utterance would allow. Then drawing himself to his full height, he turned proudly upon his heel, and walked away in the direction of his own domicil, *as straight as an Indian*, nor deigned to look behind while in sight of the tavern. The gentleman with more money than brains, was for once lost in astonishment, and longer motionless and silent than he had ever been before." [Footnote: Col. Stone.]

He held the mere sensualist in equal contempt. "Many years ago, before the Indian towns were broken up along the valley of the Genesee, a clan of the Senecas resided at Canawangus, in the vicinity of the present town of Avon. The chief of the clan was a good, easy man, named Hot Bread. He was a hereditary sachem, not having risen by merit, was weak and inefficient, and of gluttonous habits. On a certain occasion, when Mr. George Hosmer was accompanying Red Jacket to an Indian council, in the course of general conversation he inquired the chief's opinion of Hot Bread. 'Waugh!' exclaimed Red Jacket: 'He has a little place at Canawangus, big enough for him. *Big man here,'* laying his left hand on his abdomen, *'But very small here,'* bringing the palm of his right hand *with significant* emphasis to his forehead."[Footnote: Ib.]

He loved to hold communion with the sublime and grand in nature. He never wearied when viewing the falls of Niagara, and their roar, the baritone of nature's anthem, stirred within, depths that other harmonies failed to reach. When Mr. Catlin, the celebrated Indian portrait painter, desired to obtain the orator's picture, his consent was given, but he must be represented as standing on Table Rock, "for," said he, "when I pass to the other world, my spirit will come back, and that is the place around which it will linger." [Footnote: Catim's North American Indians.]

The artist gratified the orator, and represents him as standing there in the attitude of deep thought, dressed with much care in complete Indian costume, a very interesting memorial, presenting evident marks of being one of nature's noblemen.

Since then Red Jacket has gone to his grave, and this rock where he often stood and feasted his soul on sublimities unrivalled in nature, has likewise fallen, while the world, like the impetuous flood, rolls on unconscious of both.

Of the various paintings of Red Jacket, Col. Stone remarks, "The picture by Mr. Robert W. Weir, taken in 1828, at the request of Doctor John W. Francis of New York, is of far the highest order of merit, and has become the standard likeness of the last of the Seneca orators." To this is subjoined the following description from the pen of Doctor Francis, of the orator's appearance on the occasion,

"For this purpose he dressed himself in the costume which he deemed most appropriate to his character, decorated with his brilliant overcovering and belt, his tomahawk, and Washington medal.

"For the whole period of nearly two hours, on four or five successive days, he was as punctual to the arrangements of the artist, as any individual could be. He chose a large arm chair for his convenience, while his interpreter, as well as himself, was occupied for the most part in surveying the various objects, which decorated the artist's room. He had a party of several Senecas with him, who, adopting the horizontal position, in different parts of the room, regaled themselves with the fumes of tobacco, to their utmost gratification. Red Jacket occasionally united in this relaxation; but was so deeply absorbed in attention to the work of the painter, as to think, perhaps, of no other subject. At times he manifested extreme pleasure, as the outlines of the picture were filled up. The drawing of his costume, which he seemed to prize, as peculiarly appropriate, and the falls of Niagara, scenery at no great distance from his residence at the reservation, forced him to an indistinct utterance of satisfaction. When his medal appeared complete in the picture, he addressed his interpreter, accompanied by striking gestures; and when his noble front was finished, he sprang upon his feet with great alacrity, and seizing the artist by the hand, exclaimed with great energy, 'Good! Good!' The painting being finished, he parted with Mr. Weir with a satisfaction apparently equal to that which he doubtless, on some occasions had felt, on effecting an Indian treaty. Red Jacket must have been beyond his seventieth year when the painting was made. He exhibited in his countenance, somewhat of the traces of time and trial, on his constitution. Nevertheless he was of a tall, erect form, and walked with a firm gait. His characteristics are preserved by the artist to admiration; and his majestic front exhibits an attitude surpassing every other, that I have ever seen of the human skull. As a specimen for the craniologist, Red Jacket need not yield his pretensions to those of the most astute philosopher. He will long live by the painting of Weir, the poetry of Halleck, and the fame of his own deeds."

Red Jacket had a quick and acute perception, he was very adroit. He at one time exposed the false pretenses of Jemima Wilkinson by arranging it with a few Indians to converse in her presence, in a manner that excited her curiosity. The ruse was successful, she anxiously inquired what they were talking about? Turning upon her a searching glance, he exclaimed, "What! Are you Jesus Christ? and not know Indian?"

Though unacquainted with the usages of society, in the refined circles where he often appeared, he readily adapted himself to the new position, and conducted with propriety and ease, careful to conceal his ignorance at the time. Mr. Thomas Morris in a letter to Colonel Stone, observes: "He once on his return from Philadelphia, told me that when there he perceived many things, the meaning of which he did not understand, but he would not make inquiry concerning them there, because they would be imputed to his ignorance. He therefore determined on his return to ask me.

"He said when he dined at General Washington's, a man stood all the time behind his chair, and would, every now and then run off with his plate, and knife and fork, which he would immediately replace by others. 'Now,' said Red Jacket, 'what was this for?' I replied that he must have observed on the president's table a variety of dishes, that each dish was cooked in a different manner, and that the plates and knives and forks of the guests, were changed as often as they were helped from a different dish. 'Ah!' said he, 'is that it?' I replied in the affirmative. 'You must then suppose,' he continued, 'that the plates, and knives, and forks, retain the taste of the cookery?' Yes, I replied. 'Have you then,' he added, 'any method by which you can change your palates every time you change your plates? For I should suppose that the taste would remain on the palate longer than on the plate?' I replied that we were in the habit of washing that away by drinking wine. 'Ah!' said he, 'now I understand it. I was persuaded that so general a custom among you was founded in reason, and I only regret that when I was in Philadelphia I did not understand it; when dining with General Washington and your father. The moment the man went off with my plate I would have drunk wine until he brought me another; for although I am fond of eating, I am more so of drinking.'" [Footnote: Col. Stone's Life of Red Jacket.]

It has been well observed of him, "He had an innate refinement and grace of manner, that stamped him the true gentleman, because with him these virtues were inborn, and not simulated or acquired." [Footnote: W. C. Bryant's Memorial Address.]

On one occasion when Mr. George Hosmer of Avon, and several others of his tribe, were on their way to attend a certain treaty, the Indians one evening after the fatigues of the day, were unusually mirthful. Red Jacket conceiving the idea that Mr. Hosmer, who was unacquainted with their language might suppose he was the subject of their mirth, caused them to be silent, and through his interpreter, Captain Parrish, thus addressed him.

"We have been made uncomfortable by the storm; we are now warm and comfortable, it has caused us to feel cheerful and merry. But I hope our friend who is traveling with us will not feel hurt at this merriment, or suppose that we are taking advantage of his ignorance of our language, to make him in any manner the subject of our mirth."

To which Mr. Hosmer replied, that knowing himself to be in the company of brave and honorable men, he could not allow himself to entertain such an impression. After which they resumed their merriment, and Red Jacket his gravity. [Footnote: Col. Stone.]

The first efforts to construct a bridge at Niagara Falls was unsuccessful. It was supposed the force of the water where it flowed smoothly, would not be as great as where it dashed against the rocks and appeared more boisterous. This was a mistake. Every endeavor to fix a bent where the water was smooth, proved utterly abortive. At length an architect conceived the idea of placing the bridge, down where the water began to be broken in its descent, and of obtaining a foot-hold for his bent, behind some rock against which the water dashed. This resulted in the successful completion of a bridge, leading to Goat Island. After its completion, Red Jacket, in company with General Porter, was passing over it one day, when the chief, whose curiosity was excited, examined minutely every part of its construction, evidently regarding it, as a great wonder. At length discovering the secret, he exclaimed, "*Ugh! still water*!" and immediately added, "*d—n Yankee.*" [Footnote: Given to the author by T. M. Howell, Esq., of Canandaigua, N. Y.]

Red Jacket was not a stranger to *tender and refined sensibilities.*

William Savary in his Journal, while attending the Indian treaty held at Canandaigua in 1794, speaks of the children of Red Jacket in terms of high commendation. Most of them died of consumption, "in the dew of their youth."

On one occasion, when visiting an aged lady of his acquaintance near Avon, who from early life had been more or less familiar with his history, she inquired of him, if any of his children were still living? Fixing his eyes upon her, with a sorrowful expression, he replied:

"Red Jacket was once a great man, and in favor with the Great Spirit. He was a lofty pine among the smaller trees of the forest. But after years of glory he degraded himself, by drinking the firewater of the white man. The Great Spirit has looked upon him in anger, and his lightning has stripped the pine of its branches." [Footnote: Related to Col. Stone by Mrs. George Hosmer of Avon.]

Some four or five years before his death, three brothers, named Thayer, were executed at Buffalo for the crime of murder. The occasion was unusual, and multitudes of both sexes, from the surrounding region, flocked to witness the unhappy spectacle.

On the day of the execution, Red Jacket was met by Judge Walden, of Buffalo, wending his way from the town to his home. The judge inquired where he was going? At the same time expressing his surprise that he did not go with the multitudes, flocking to witness the spectacle. His answer was brief; "Fools enough there already. Battle, is the place to see men die."

The reply was a merited rebuke to the desire so prevalent, to witness these awful sights. [Footnote: Mrs. George Hosmer to Col. Stone.]

Red Jacket ever cherished a watchful regard over the interests of his people, and was always ready to speak in their behalf.

At the trial of an Indian for burglary, himself and other chiefs were present to render any aid in their power, to their brother in bonds. The prisoner was found guilty of having broken into a house and stolen a few silver spoons. The crime of petit larceny, was thus merged in the greater one of burglary.

At a fitting opportunity Red Jacket arose and spoke eloquently in his brother's defense; urging the independence of his nation, the existence among them of laws for the punishment of theft, and boldly demanding the surrender of the prisoner, assuring the court that the prisoner should be tried by these laws, and suffer the penalty they demanded. His effort though regarded as able and brilliant, did not avail to rescue the prisoner from the white man, whose sentence in the case being for burglary instead of theft, Red Jacket regarded as unnecessarily severe.

When the proceedings were over, Red Jacket, who happened to be standing with a group of lawyers, took the following method of expressing his dissatisfaction.

Beholding on the sign of a printing office near by, an emblematic representation in large figures and characters, of Liberty and Justice; he asked in broken English, pointing to one of them, "*What-him-call?*" It was answered, *Liberty*. "Ugh!" was his significant and truly aboriginal response. Pointing then to the other figure, he inquired, "*What*-HIM-call?" It was answered, JUSTICE. Whereupon his eye kindling with animation, he asked with evident emotion, "WHERE-HIM-LIVE- NOW?" [Footnote: Geo. Hosmer, Esq., to Col. Stone.]

If the sincerity of Red Jacket's regard for the welfare of his people was ever questioned, it was by those who knew not his inner self. In guarding the interests of his people, he was in the habit of closely watching strangers, not only, but even his own friends.

Owing to slanderous reports that had been circulated, he at one time began to suspect that his friend Captain Jones, was actuated by motives of self- interest, and did not property regard the interest of the Indians.

Jones soon after met Red Jacket with his usual cordiality of manner, but was received with evident marks of coldness and distrust. "After the lapse of a few minutes, during which time the questions of Jones were answered in monosylables, the captain asked an explanation of the orator's conduct. Fixing his searching glance upon him, as if reading the secrets of his soul, Red Jacket told him of the rumor circulated, in reference to his fidelity to the

Indians, and concluded by saying with a saddened expression, 'And have *you* at last deserted us?' The look, the tone, the attitude of the orator, were so touching, so despairing, that Jones, though made of stern materials, wept like a child; at the same time refuting the calumny in the most energetic terms. Convinced that Jones was still true, the chief, forgetful of the stoicism of his race, mingled his tears with those of Jones, and embracing him with the cordiality of old, the reconciled parties renewed old friendship over a social glass." [Footnote: W. H. C. Hosmer to Col. Stone.]

CHAPTER XXI

Views at the close of life—Incident—His life work—Unfavorable influences—Advance of the Christian party—Conversion of Red Jacket's wife—He leaves her—His return—Red Jacket deposed—Journey to Washington —His restoration—Rapid decline—Regards his end as near—Talks with his people—Endeavors to unite them.

With the views entertained by Red Jacket, the objects that met him on every side, as he drew near the close of life, were far from pleasant. Yonder hillside, exposed to the gaze of the world, its huge rocks laid bare; those fields, stretching further than eye could reach, bounded not by woodland, lake, or river, but by the white man's fence; ten thousand dwellings, smiling with the abundance and thrift of the husbandman, city and village, bustling with tumult, and the noise of busy hammers, and rattling wheels, and roaring engines; all of these however gratifying to the white man, as marks of improvement, afforded him no pleasure. He saw in them the sepulcher of his people's pride and glory.

The hillside opened to the sunlight, for the innocent lamb to sport upon, or to make the stable ox a home, he would have loved better, as when sheltered once by the sturdy oak or stately pine, its rocks jutting out from behind the ivy, and its bosom threaded by the path of the deer. The fields might have appeared inviting and green, but the white man's barrier would have warned him away, the road he would have looked upon as a prisoned path, and he would have taken to the woods, as a place more congenial to his spirit.

It is said of him "that in the days of his youth he was wont to join the hunters in the beautiful valley of the Genesee, with great enthusiasm. Game was then plenty, and they were the finest hunting grounds, he could traverse. Toward the close of his life he went thither to indulge once more, in the pleasures of the chase, where a forest apparently of considerable extent, yet remained. He entered it, recognizing some of his ancient friends among the more venerable of the trees, and hoping yet to find abundant game. But he had not proceeded far before he approached an opening; and his course was presently impeded by a fence, within the enclosure of which, one of the pale faces was guiding the plow. With a heavy heart he turned in another direction, the forest seeming yet to be deep, and where he hoped to find a deer, as in the days when he was young. But he had not traveled long, before another opening broke upon his view, another fence impeded his course, and another cultivated field appeared within. He sat down and wept." [Footnote: Circumstances related to Col. Stone by a Seneca chief.]

It has been well observed: "The whole life of the Seneca chief was spent in vain endeavors to preserve the independence of his tribe, and in active opposition as well to the plans of civilization proposed by the benevolent, as to the attempts at encroachment on the part of the mercenary.... He yielded nothing to persuasion, to bribery, or to menace, and never to his last hour remitted his exertions, in what he regarded the noblest purpose of his life." [Footnote: McKenney's Biography.]

But at the close of life, Red Jacket began to realize more than ever the power of those forces bearing down upon him, to resist which he had summoned all the energies he could command. His people, notwithstanding his efforts, were constantly brought by the encroachments of the whites, into a narrower compass, and the religion and customs of the whites continued to gain ground, and threatened to supercede the time honored usages of his fathers.

Intoxicating drinks also, the bane of the Indian race, wrought sad havoc among his people, and had well nigh ruined himself. His influence was thus effectually crippled, and his opposition to Christianity, and the efforts of the whites to obtain their land, carried much less weight, than at an earlier period of his life. He saw and felt this, and in view of it, was much cast down.

His opposition to Christianity, is said to have been much encouraged by wicked and designing men among the whites, who feared that the presence of missionaries among the Indians, would interfere with their unworthy and base designs.

But his decision when formed, as already intimated, was consistently and perseveringly maintained. He narrowly watched every proceeding, gathered around him such as would be controlled by his influence, or example, and inculcated in them those sentiments of steadfastness, in the religion of their fathers, so strikingly manifested in his own conduct.

After various discouragements and reverses, the missionary was at length established among his people, and the adherents of Red Jacket, which at first were the most numerous, by degrees diminished, until finally those friendly to Christianity, outnumbered the others. Red Jacket's people one by one, became interested in the religion the missionary had come to teach. The schools established began to be well attended, several chiefs embraced the new religion; some of them were men of influence and carried with them many others. Finally in 1826, Red Jacket's wife became interested on the subject of religion, attended the meetings of the Christians, was led to abandon the pagan worship, she formerly attended, altogether, and giving evidences of piety, proposed to unite with the mission church, under the care of the Rev. Mr. Harris.

Before uniting, she laid the subject before Red Jacket and desired his consent. This he utterly refused, and threatened in case she did so, to leave her and never visit her again. Her trial in view of this refusal, she referred to Mr. Harris, who kindly endeavored to show her what the law of Christianity demanded, that it required her to obey God, rather than man; that though her course might subject her to trial, she had the promise of the grace of Christ to help her, and that in the end it might promote her good. Still he committed the matter wholly to her own conscience, advising her to pursue the course that might thus be indicated, and leave the event in the hands of God.

After deliberating for a time she united with the church, and Red Jacket, true to his threat, left her and went to another reservation. She bore his displeasure with a meek and Christian spirit, remained at home with her family, and conducted discreetly, pursuing as before the duties of her household.

Red Jacket after a few months' absence returned, desiring to be welcomed again by his wife, who received him on condition of his not interfering with her, in her religious views, or attendance on the meetings of the mission. To this he gave his assent, and was ever afterward faithful in observing his pledge; not opposing, but aiding her in performing, according to her desire, her religious obligations.

A division was now apparent among the Senecas, in regard to religion. There was a Christian, and a Pagan party. The former led by Young King, Captain Pollard, and others; the latter recognized Red Jacket as its ruling spirit.

The opposition he had so long exerted, began to be regarded with impatience. As the Christian party advanced and became more numerous, they were unwilling to submit to the dictation of the orator. They began to feel that in his opposition to the education and improvement of his people, he was acting the part of an enemy, and not a friend.

His habits of intemperance also, having greatly lessened their esteem, they became unwilling he should longer hold the commanding position he had enjoyed, and so well adorned, in the earlier part of his life. At a council held in September, 1827, a paper was drawn up, containing charges against the orator, which were assigned as a reason for the extraordinary course they pursued, closing with the declaration, that they renounced him as their chief, and forbade him to act as one, affirming that he should thereafter be regarded as a private man.

This proceeding stung the orator to the quick, and aroused him to action, He could not endure the thought of the humiliation thus brought upon him, at the close of life. The thought too, that it had been effected by those who differed from him, in their religious sentiments, and would be regarded as a triumph over him, touching the views he had long entertained, as to what would best promote the welfare of his people, affected him in a point so near his heart, as to forbid his resting under it.

"It shall not be said," thought he, "that Sa-go-ye-wat-ha, lived in insignificance, and died in disgrace. Am I not yet strong? Have I not yet power to withstand my enemies?"

He set out for Washington, to spread his griefs before his great Father. On arriving there he visited Colonel McKenney, who had charge of Indian affairs. That officer had been informed, through the Indian agent, of all that had transpired among the Senecas, and of the cause of their displeasing Red Jacket.

When the customary salutations were over, Red Jacket remarked through his interpreter, "I have a talk for my Father." "Tell him," said Colonel McKenney, "I have one for him. I will make it, and will then listen to him." The colonel then proceeded to give a minute history of all that had recently transpired, and dwelt upon the various causes that had operated in producing the rupture, that had taken place. He pointed out to him the course he ought to have pursued, that he should have manifested a spirit of forbearance, and allowed the Christian party the same liberty in the exercise of their sentiments, which he demanded for himself; and that this course would have saved him the mortification he now experienced, in being expelled from office and power.

During this conversation Red Jacket never took his keen and searching eye from the speaker, but at its close turned to the interpreter, and pointing in the direction of his home and people, said, "*Our Father has got a long eye.*"

He then proceeded to vindicate himself and his cause, not forgetting to pour upon the Black coats plentiful effusions of wrath. The colonel advised him to return to his people, convene a council and come to a better understanding with them, by allowing those among them who desired to do so, to become Christians, while himself and those who thought like him, might claim the privilege of following unmolested, the faith of their fathers. [Footnote: Col. McKenney's Indian Biography.]

About one month had passed since Red Jacket's deposition. In the mean time Red Jacket had been very active in going from one reservation to another, and sparing no pains, in gathering a Great Council, from those belonging to the Six Nations.

Another council was convened, much larger than the former, composed of members from other reservations, belonging to the Iroquois confederacy. It assembled at the upper council-house of the Seneca village near Buffalo.

At the opening of the council, the paper declaring the orator's deposition was read. Half Town, a Seneca chief of the Cattaraugus reservation then arose, and said there was but one voice in his nation, and that was of general indignation at the contumely cast on so great a man as Red Jacket. The council was then addressed by several other chiefs very much to the same effect. After which the condemned orator arose slowly, as if grieved and humiliated, but yet with his ancient air of command.

"My Brothers:" said he, after a solemn pause, "You have this day been correctly informed of an attempt to make me sit down, and throw off the authority of a chief, by twenty-six misguided chiefs of my nation. You have heard the statements of my associates in council, and their explanations of the foolish charges brought against me. I have taken the legal and proper way to meet these charges. It is the only way in which I could notice them. Charges which I despise, and which nothing would induce me to notice, but the concern which many respected chiefs of my nation, feel in the character of their aged comrade. Were it otherwise I should not be before you. I would fold my arms, and sit quietly under these ridiculous slanders.

"The Christian party have not even proceeded legally to put me down." He then made some artful observations on the origin of the attack made upon him. He laid open its history step by step. He dwelt upon the various circumstances connected with the introduction of Christianity among them. He alluded to the course taken by the Christians as ruinous and disgraceful, especially in their abandonment of the religion of their fathers, and their sacrifices, and of the lands given them by the Great Spirit, for paltry considerations. As for the *Black coats*, Mr. Calhoun had told him at Washington four years before, that the Indians must treat with them as they thought proper; the government would not interfere. "I will not consent," said he, sagaciously identifying his disgrace with his opposition to the Christians, "I will not consent silently to be trampled under foot. As long as I can raise my voice, I will oppose such measures. As long as I can stand in my moccasins, I will do all I can for my nation. Ah! it grieves my heart, when I look around me and see the situation of my people, in old times united and powerful, now divided and feeble. I feel sorry for my nation. Many years have I guided my people. When I am gone to the other world, when the Great Spirit calls me away, who among them can take my place?" [Footnote: Thatcher's Indian Biography.]

No adequate account of this speech has been preserved. It is said he spoke three hours in his own defense; that it was a masterly effort, and equal to the speeches he used to make in his palmiest days. [Footnote: Conversation of the author with Wm. Jones, Seneca chief.]

Though greatly dilapidated in his powers by intemperance, he was thoroughly aroused on this occasion, and the eloquence, pathos, and fire of a former day, shed around him the luster of a superior mind, and his people for the time, forgot and forgave his delinquencies, and by unanimous consent, reinstated him in office and power.

Thus by means of one more great exertion of this wonderful faculty, by which he controlled the minds of his people, they were led to reverse the decision that had been made against him, and though he stood among them but the blasted trunk of that tree, which, in its full and luxuriant prime, cast a deep and mellowing shade over their closing history, and invested it still with the appearance of strength; they resolved he should yet wear the title, that better befitted him in other days, though it served but slightly to hide the deformity, wrought in his noble nature, by the demon of intemperance.

With this speech the public career of Red Jacket is closed. The effort he made on this occasion, added to his exertions previous to the gathering of the council, was too great for his aged and enfeebled condition. After this he declined very rapidly, and seemed to realize that his end was drawing near. He often adverted to this event, but always in language of philosophic calmness.

In view of it he visited successively all of his most intimate friends, at their cabins, and talked with them in the most impressive and affecting manner. He told them that he was passing away, and his counsels would soon be heard no more. He ran over the history of his people, from the most remote period to which his knowledge extended, and pointed out as few could, the wrongs, the privations, and the loss of character, which almost of themselves constituted that history. "I am about to leave you," said he, "and when I am gone, and my warnings shall be no longer heard, or regarded, the craft and avarice of the white man will prevail. Many winters have I breasted the storm, but I am an aged tree, I can stand no longer. My leaves are fallen, my branches are withered, and I am shaken by every breeze. Soon my aged trunk will be prostrate, and the foot of the exulting foe of the Indian, may be placed upon it in safety; for I leave none who will be able to avenge such an indignity. Think not I mourn for myself. I go to join the spirits of my fathers, where age cannot come; but my heart fails, when I think of my people, who are soon to be scattered and forgotten."

Many noticed that his feelings at this time were greatly modified and mellowed, with respect to the stand he had taken against Christianity. His wife's example, who was a woman of humble, consistent piety, exerted a salutary, and happy influence upon him. It led him to regard Christianity more favorably, and to recede very much from the hostile position he had previously maintained. He talked of peace, and sought to bring about a reconciliation between the two parties. He convened a council with this in view. He made special preparations to attend it, dressing himself with more than ordinary care, with all his gay apparel and ornaments. He went with the intention of making what would have been his farewell speech, and giving them his last counsel.

He was taken suddenly ill at the Council-house, of cholera morbus and returned home, saying to his wife, "I am sick; I could not stay at the council, I shall never recover."

He then took off his rich costume, and laid it carefully away, reclined upon his couch, and did not rise again till morning. His wife prepared him medicine, which he took, but said, "it will do no good. I shall die."

The next day he called his wife and the little girl he loved so much, requested them to sit beside him and listen to his parting words. Addressing his wife, he said: "I am going to die, I shall never again leave this house alive. I wish to thank you for your kindness to me. You have loved me. You have always prepared my food, and taken care of my clothes, and been patient with me. I am sorry I ever treated you unkindly. I am sorry I left you, because of your new religion. I am convinced it is a good religion, and has made you a better woman, and wish you to persevere in it. I should like to live longer for your sake. I meant to build you a new house, and make you more comfortable, but it is now too late."

Addressing his daughter, he said; "I hope my daughter will remember what I have so often told her, not to go in the streets with strangers, or associate with improper persons. She must stay with her mother, and grow up a respectable woman."

He said again: "When I am dead, it will be noised abroad through all the world, they will hear of it across the great waters, and say, Red Jacket the great orator is dead. And white men will come and ask you for my body. They will wish to bury me. But do not let them take me. Clothe me in my simplest dress, put on my leggins and my moccasins, and hang the cross I have worn so long, around my neck, and let it lie upon my bosom. Then bury me among my people. Neither do I wish to be buried with Pagan rites. I wish the ceremonies to be as you like, according to the customs of your new religion, if you choose. Your minister says the dead will rise. Perhaps they will. If they do, I wish to rise with my old comrades. I do not wish to rise among pale faces. I wish to be surrounded by red men. Do not make a feast according to the customs of the Indians. Whenever my friends chose, they could come and feast with me, when I was well, and I do not wish those who have never eaten with me in my cabin, to surfeit at my funeral feast."

When he had finished he laid down on his couch and did not rise again. He lived several days but was most of the time in a stupor, or else delirious. He often asked for Mr. Harris, the missionary, and would afterward unconsciously mutter: "I do not hate him. He thinks I hate him, but I do not, I would not hurt him." The missionary was sent for repeatedly, but was from home at the time, and did not return till after the chief's death.

When the messenger told him Mr. Harris had not come, he replied: "Very well, the Great Spirit will order it as he sees best, whether I shall speak with him or not." [Footnote: Conversation of the author with Wm. Jones, Seneca chief, and sketch of Red Jacket in "The Iroquois." The account of the orator's closing hours given in this work, is more full, but in perfect accord with the statements made to the author by Mr. Jones.] Again he would murmur: "He accused me of being a snake, and trying to bite somebody. This was true, and I wish to make satisfaction."

The cross he wore was a very rich one of stones set in gold, and large; by whom it was given, his friends never knew. This is all the ornament he requested to have buried with him.

It was customary among the Indians to make funeral feasts. No family was so poor as not thus to honor the dead. If all they possessed was a cow, it was slaughtered for the occasion. Red Jacket desired nothing of this kind. A pagan funeral for a distinguished person is a pompous affair, and lasts for ten days. Every night a fire is kindled at the grave, and around it the mourners gather, and utter piteous wails.

The wife and daughter were the only ones to whom he spoke parting words, or gave a parting blessing. As his last hour drew nigh, his family all gathered around him, but the children were not his own, they were step-children, his own were all sleeping in the churchyard, where he was soon to be laid.

His step-children he always loved and cherished, their mother had taught them to love and honor him. The wife sat by his pillow and rested her hand on his head. At his feet stood the two sons, now aged and Christian men, and by his side the little girl, whose hand rested on his withered and trembling palm. His last words were still, "Where is the missionary?" He then clasped the child to his bosom, while she was sobbing in anguish, her ears caught his hurried breathing, his arms relaxed their hold, she looked up, he was gone.

There was mourning in the household, there was great mourning among the people. The orator, the man of matchless gifts, of surpassing eloquence was no more; and there were none to fill his place.

Red Jacket desired after his death, a vial of cold water might be placed in his hand. His reason for this his friends did not understand. Red Jacket felt that intemperance had been the bane of his life. Possibly from this conviction he may have desired to be accompanied in his journey to the spirit-land, by the beverage of which his better judgment most approved.

The arrangements of his funeral Red Jacket committed to his wife's son-in- law Wm. Jones. His friends, who belonged mostly to the Christian party, chose to have at his funeral the simple and appropriate services of that religion. It was largely attended by his own race, and by the whites living in that vicinity. He was buried in the mission burying ground, where were reposing many of his race, the aged and young, warrior, sachem, child.

His death was at his residence near the church and mission-house at Seneca village on the 20th of January, 1830.

INDEX